New Updated Edition

REMOVING
obstacles

to SAFETY
A BEHAVIOR-BASED APPROACH

New Updated Edition

REMOVING
obstacles

to SAFETY
A BEHAVIOR-BASED APPROACH

Judy Agnew & Gail Snyder

PMP

Performance Management Publications (PMP)

PMP

Performance Management Publications (PMP)
3344 Peachtree Road NE, Suite 1050
Atlanta, GA 30326
678.904.6140

ISBN-10: 0-937100-07-2
ISBN-13: 978-0-937100-07-3
Library of Congress Control Number: 2002106180

Printed in the United States of America
ViaTech Publishing Solutions

3 4 5 6 7

Lisa Smith, Art Director (cover and text design)

PMP is a division of Aubrey Daniels International, Inc.

PMP books are available at special discounts for bulk purchases by corporations, institutions, and other organizations. For more information, please call 678.904.6140, ext. 131 or e-mail lglass@aubreydaniels.com

DEDICATION

I dedicate this book to the clients and colleagues I have had the privilege of working with over the years, and to my family, Bruce, Matthew, and Kianna. I have learned about behavior from all of you.

– Judy Agnew

I dedicate this book to my husband, Jack, and son, David, for their loving support, cooperation, and willingness to allow me hours of uninterrupted writing.

– Gail Snyder

ABOUT THE AUTHORS

Judy L. Agnew, Ph.D., is Vice President for Aubrey Daniels International and specializes in designing behavior-based business solutions. With more than 17 years of consulting experience and a Ph.D. in Applied Behavior Analysis, she works with her clients to develop customized interventions that are well grounded in the science of behavior. Dr. Agnew has worked in many industries including oil and gas, mining, forest products, utilities, manufacturing, distribution, assembly, insurance, banking, newspapers, retail, and construction. In addition to her consulting work, she has applied her organizational and creative skills to instructional design projects including the development of core curricula for ADI Train the Trainer courses in both Performance Management and Behavior-Based Safety.

Gail Snyder received her undergraduate degree in journalism and her master's in communications from Georgia State University. As a former editor of *Performance Management Magazine*, she specialized in making scientific principles and research enjoyable and understandable to a general audience. Over the past twenty years she has edited and written articles and books focusing on corporate management strategies. Currently she works as an Atlanta-based freelance writer and editor.

TABLE OF CONTENTS

NOTE TO OUR READERS

This book is intended to be an easy-to-read introduction to behavior-based safety (BBS). Several good books are available that provide more information on Applied Behavior Analysis (the science upon which BBS is founded) and philosophical discussions about safety and behavior. Our intent is to offer potential users of this process an introduction to the basics. In this book, you won't find every detail required to implement BBS, but you will find a discussion of the key components necessary to make BBS work. While this book refers to behavior-based safety in a generic way, the case studies and the methods presented are based on Aubrey Daniels International's (ADI) unique approach to BBS and years of implementing that approach successfully in hundreds of companies.

We hope this book provides you with the fundamental understanding you need to begin using this powerful technology called behavior-based safety.

PART I

WHY TAKE A BEHAVIORAL APPROACH TO SAFETY?

CHAPTER 1

WHY ARE PEOPLE STILL GETTING HURT ON THE JOB?

A laborer removing an internal stairway falls 32 feet to his death. A railroad veteran of 26 years is crushed to death between two train cars. A construction worker suffocates when the dirt walls of a pipeline trench collapse. These are the extreme results of on-the-job accidents, the types of incidents that occasionally reach the daily news.

Yet, when accessing OSHA's Web site (www.osha.gov) one can find a lengthy list of the types of work-related accidents that have resulted in death. Each report represents only one category of fatal accident, ranging from falls to explosions to electrocutions. For example, in the most recent data available, of the more than 5000 people killed at work in America that year, 42 percent were killed in transportation incidents, 27 percent by contacts with objects and equipment, 14 percent in falls, 13 percent by workplace violence, and 4 percent in fires and explosions.

OSHA not only reports the annual statistics of how many workers have lost their lives while simply trying to earn a living, the agency also monitors the on-the-job incidents and injuries that occur daily throughout the United States. Ultimately, these on-the-job incidents, injuries, and accidental deaths are tabulated, categorized, and reduced to numerical blips on statistical records. But, before such statistics are compiled, the cause of every incident must first be discovered.

After a reportable on-the-job injury, the discovery process inevitably involves claims of one or all of the following:

- The organization didn't provide a safe environment.
- The safety process wasn't adequate or wasn't communicated clearly.
- The equipment was faulty.
- Management didn't support the safety process.
- Somebody didn't know and/or didn't practice safe behavior.

Though any or even all of the above statements may apply, only one factor—behavior—is the universal root of all incidents. Think about it!

The organization may provide a safe environment by investing in the best architects, engineers, technicians, and safety specialists—*behavior*. Someone up the line may decide that certain optional measures are too expensive and veto funding—*behavior*. The safety process might cover every conceivable rule, but the safety manager may place the procedures manual in his/her office drawer—*behavior*.

A cost analysis specialist may advise corporate executives that in-depth training for every employee is cost prohibitive—*behavior*. The executives cut the training budget—*behavior*. A purchasing executive may opt for the lowest bid, buying equipment that carries the minimum required safety features—*behavior*. That same purchasing executive may choose the safest equipment available, only to have the safety features bypassed or even removed by those who use the equipment—*behavior*. Management may talk of safety support, but, through their actions and words indicate to employees that when the rubber hits the road, production and getting the job done on time takes first priority—*behavior*.

Get the picture?

Place, process, equipment, management support, and personal safety depend on safety leadership at all levels and safety leadership requires safe behavior at all levels. "Rules and regulations are necessary for safety," stated Jeffrey L. Feerer, Ph.D., environmental, health, and safety leader for Dow Chemical. "But we have a sterling compliance record with OSHA and people still get hurt. It's because of the behavioral component. People are still doing things that result in injury."

Doing things safely is the core of behavior-based safety. The essential elements of the entire process consist of pinpointing specific safe behaviors, observing and collecting data on how often those behaviors are performed, providing feedback to the group(s) of performers, and recognizing and rewarding improvement. BBS is not about accusation, punishment, or absolving management of accountability. With BBS, safe behavior becomes the responsibility of every person in an organization, including those whose decisions and actions impact safety and the safety process. Rules and regulations are a crucial first step to a safe environment, but a system that makes people *want* to comply with those rules is the key to safety.

Why Is Behavior-Based Safety Necessary?

Safety can be divided into at least three different areas. First is the physical environment, things like lighting, flooring, guards on machines, availability of personal protective equipment (PPE), and so on. Second is a safe work process; that is, how to do the work safely. This includes training and procedures. Safe-lifting videos and lockout/tag-out procedures fall into this category. Even if these first two areas were perfect (which they never are), companies would still have accidents. Why? Because even when the physical environment is very safe, procedures exist to address all potentially dangerous activities, and everyone has up-to-date, effective training in safety, people still get hurt. Having personal protective equipment available and procedures that tell people when and how to use that equipment does not guarantee that people *will* use it. Having guards on machines doesn't mean people will use the guards. People can still bypass guards, take shortcuts, and slip back into at-risk habits. We must add a focus on behaving safely (doing the right things) such as using the guards, following procedures, and so on. BBS focuses on ensuring that people do the right thing on the job whether they are frontline employees doing things that directly affect their own safety or managers and executives making decisions that affect the safety of others. Please take a look at the figure on the next page taken from OSHA's Web site.

ACCIDENT REPORT

FATAL FACTS

U.S. Department of Labor
Occupational Safety
and Health Administration
No. 18

ACCIDENT SUMMARY

Accident Type	Caught by Rotating Part
Weather	Clear
Type of Operation	Telephone Line Installation
Crew Size	3
Collective Bargaining	No
Competent Safety Monitor on Site?	Yes
Safety and Health Program in Effect?	Yes
Was the Worksite Inspected Regularly?	Yes
Training and Education Provided?	No
Employee Job Title	Boring Machine Operator
Age/Sex	56/M
Experience at this Type of Work	10 years
Time on Project	5 days

BRIEF DESCRIPTION OF ACCIDENT

A three-man crew was installing an underground telephone cable in a residential area. They had just completed a bore hole under a driveway using a horizontal boring machine. The bore hole rod had been removed from the hole. While the rod was still rotating, the operator straddled it and stooped over to pick it up. His trouser leg became entangled in the rotating rod and he was flipped over. He struck tools and materials, sustaining fatal injuries.

INSPECTION RESULTS

Following its inspection, OSHA issued one citation for one alleged serious violation of its construction standards. Had the equipment been properly guarded, this fatality might have been prevented.

ACCIDENT PREVENTION RECOMMENDATIONS

1. Employees must be instructed to recognize and avoid unsafe conditions associated with their work (29.CFR 1926.21 (b)(2)).
2. Guards must be installed on moving parts of equipment with which employees may come into contact (29 CFR 1926.300 (b)(2)).

SOURCES OF HELP

- Construction Safety and Health Standards (OSHA 2207) which maintains all OSHA job safety and health rules and regulations (1926 and 1910) coveting construction.
- OSHA-funded free consultation services. Consult your telephone directory for the number of your local OSHA area or regional office for further assistance and advice (listed under U.S. Labor Department or under the state government section where states administer their own OSHA programs).

NOTE: The case here described was selected as being representative of fatalities caused by improper work practices. No special emphasis or priority is implied nor is the case necessarily a recent occurrence. The legal aspects of the incident have been resolved, and the case is now closed.

Courtesy of U.S. Department of Labor, Occupational Health and Safety Administration

Notice the age and experience level of the employee who was killed in this fatal tragedy. If you access this Web site at www.osha.gov and search for Fatal Facts, you will discover a listing of fatal accident categories, each category representing only one typical example of a type of fatal injury that may have occurred many times over.

FATAL ACCIDENT CATEGORIES

Struck by Nail

Fall from Different Level

Explosion

Struck by Collapsing
Crane Boom

Caught in Between

Fall from Elevation

Crushed by Falling Wall

Struck by Falling Object

Trench Cave-in

Crushed by Falling
Machinery

Electrocution

Collapse of Shoring

Caught by Rotating Part

Fire/Explosion

Asphyxiation

Electrical Shock

Fall and Drowning

Death Due to Burns

Fall from Roof

Fall from Excavator
Bucket

Keep in mind that the fatal facts listing above only describes incidents "selected as representative of fatalities caused by improper work practices" that resulted in death. The details of these cases show that even years of experience do not always prevent a one-time, at-risk behavior and if that risky behavior fails to pay off only one time, it may be the last time. For example, one national news story highlighted the luck of a construction worker who was the accidental recipient of a nail shot through his skull via a coworker's nail gun. X-rays showed a 3-inch nail lodged in his brain that had missed all vital nerve and functional areas. The surgeon in this case called the worker "the luckiest man on Earth." On the other hand, take a look at the first item on the Fatal Accident Categories. Safety and on-the-job survival should not be a fluke!

These are the details for a very small sample of fatal on-the-job accidents that are repeated every day in the United States. And, this OSHA listing doesn't even include on-the-job lost-time injuries and permanent disabilities. Clearly, organizations must spell out safety procedures and recommendations and provide training, but then the major variable that remains is whether or not those procedures are followed.

Summary

- Because behavior is such an important part of safety, addressing behavior in a systematic way is a critical part of any safety system.

- The word *behavior* in behavior-based safety does not only refer to the behavior of frontline workers. BBS should address behavior at all levels of the organization.

Did You Know?

- Men are the victims of more than 90 percent of all job-related accidents.

- Accidental deaths appear much less significant a risk compared to cancer and heart disease when you just count the deaths caused by same. However, if you compare the number of lost years, accidents jump into first place among the killers of Americans.

- The National Center for Health Statistics estimates that in a 100-person office, the most frequent reasons given for taking sick days and the average number taken per year are colds, fractures, sprains, and flu.

- If we had the same mortality rate now as in 1900, more than half the people in the world today would not be alive.

From Richard Hawk and Company, "Safety Stuff" Newsletter. Reprinted by permission of the author. www.makesafetyfun.com

CASE STUDY

Dow and Local United Steelworkers Union Partner to Promote Safety

Traditionally, behavioral safety processes are a hard sell to unions. One of the primary reasons for resistance is that many union leaders view behavioral approaches as an effort to abdicate management from any responsibility for safety. "Typically, a union looks at behavior-based safety as a means of saying at the end of the day it is always the fault of the employee that there is an unsafe situation," said Jeffrey L. Feerer, Ph.D., environmental, health, and safety leader for Dow Chemical's Midland, Michigan, plant. "Our local union agreed that this behavior-based safety process was not a fault-finding exercise. It is instead making sure that you do the safe behaviors that keep you from getting hurt."

Feerer credits the success of BBS at Midland to a grassroots implementation approach. Rather than mandate use of the BBS technology at the facility, the employees were offered training on a voluntary basis. "We agree with the union leadership that this is a process that has been proven to keep people safe, so it's something we should incorporate in the collective bargaining agreement," Feerer said.

Back to Behavior

Midland's employees manufacture and package hundreds of specialty chemicals and agricultural products. Safety, of course, is the highest priority, and Dow takes

every precaution to ensure safe equipment, environment, and procedures. Doing so has enabled the
Midland operations to claim an admirable safety record.
However, analysis of the data indicated additional
opportunities for improvements at the behavioral level.

Coaches and Peer Training

Dow trained several coaches from the union leadership
ranks, who then trained other operations personnel.
"Rather than just the managers knowing about BBS, a
significant portion of the workforce now has serious
training in the process," Feerer commented.

After training, many of Midland's work groups
formed safety teams, and then decided on three safe
behaviors to bring to habit level through peer observation, data collection, feedback, and positive recognition
and reinforcement. Also, every six months Dow asks for
volunteers from the unionized hourly ranks to train as
safety coaches. The safety coaches serve for six months,
learning about behavioral safety and talking to members of different plants about the process. "We are on
our third group of safety coaches," Feerer explained. "In
this way, we're educating the workforce and seeding
them with a whole population that knows a lot about
behavioral safety. We are starting to see the rewards."

Results Worth Celebrating

Before BBS, the plant's OSHA incident rate had leveled
off at 4.0 where it stubbornly remained for five years.
Early in the year, Dow set a global corporate goal to cut

the entire company's injury rate by 90 percent within six years. Midland set a 30 percent reduction in injury rate goal by year-end. They soon had to rethink those plans because they met the year-end 30 percent reduction goal within five months, cutting the incident rate to 2.8! "We thought it would take an entire year to come that far," Feerer said. "We held a giant picnic for all 6000 people here. It was amazing. People volunteered to cook hamburgers and hotdogs. It was a great event, working together to pull off something like that. The spirit of working together is the key to reaching the 90 percent goal."

The Components of Success

The people at Dow Midland reached the point where they realized that focusing on safe behavior, at all levels, was the logical next step. Doing so meant not only that individuals had to embrace safe behavior as a personal responsibility but also that management had to loosen its grasp on the decisions and actions surrounding safety. In turn, union leadership and personnel had to realize that individual responsibility did not equate to blame, nor did it release management from responsible behavior. A few false starts were inevitable before the balance that results in mutual cooperation and trust was achieved. Dow Midland now knows that safety is every person's business. Feerer concluded, "The biggest change is that the safety program has always been seen as the management's program. If a group of employees were mad at management, they wouldn't do the safety program because they wanted to get back at the company.

Now, safety is not management's process; it's every-body's process. It is voluntary, but we are promoting this process to use because it works. We're now finding that people are doing the process because they want to do it; and our safety rate keeps getting better and better."

CHAPTER 2

ADDING BEHAVIOR TO THE SAFETY EQUATION

Why are companies that care doing more than simply putting up safety slogans on the wall? Why are organizations suddenly realizing that behavior is an important part of the safety equation? Many companies think they have done all they can to provide safe working conditions, policies, procedures, and training; yet the efforts have resulted in a leveling of results, meaning that even if incident rates are lower, incidents still occur. This is a very common and frustrating experience for organizations. Therefore, companies are seeking opportunities to improve in these and other safety-related areas.

Does showing a safety video once really constitute good training? Do infrequent audits catch safety issues soon enough? Does the process for correcting at-risk conditions work as quickly as it could? Even if people are trained and have the right equipment, do they all follow the procedures learned in training and/or use the safety equipment provided? If the answer is *no* to any or all of these questions, the solution can be found in behavior-based safety (BBS).

Keep in mind that *all* components of your safety system are necessary. It is critical that timely audits and inspections are done, effective safety training is given, and safety procedures are written and communicated. It is also critical to ensure those procedures are followed and that behaviors identified in training actually occur. BBS builds on what you are already doing in safety by providing a *systematic* and *positive* way to ensure that everyone does his or her part to build a safe workplace through individual and group behavior, including the crucial support behaviors of management.

Objectives of BBS:

1. Increase the number of safe behaviors of all employees. In fact, the goal of a BBS process is to create safe habits. A safe habit is a safe behavior that you do consistently. In addition, a safe habit is something you do without having to stop and think about it. A common example of a safe habit is buckling the seat belt before driving. Many people buckle their seat belts 100 percent of the time before driving and they do so without thinking about it. The great thing about a safe habit is that no matter what else is going on, you do the right thing. You might be talking to someone when you get in the car, but the seat belt still gets buckled. You might be deep in thought about something that just happened to you, but the seat belt gets

buckled. So with safe habits you do the safe thing consistently and automatically. Now, imagine your work environment if your team identified all the behaviors that put you and your coworkers at risk, then together turned those behaviors into safe habits so that everyone did the safe thing every time. Imagine how many accidents could be prevented. This is the goal of the BBS process.

2. Obviously the ultimate goal of BBS is to reduce and ideally eliminate injuries. The goal is to keep everyone safe and to allow every individual to perform the behaviors that ensure personal safety.

3. To keep everyone safe, people must create safe habits at the frontline level where people have the greatest chance of getting hurt, but we also need to make sure management and other employees live up to their responsibilities to create and maintain a safe workplace. Thus, a critical objective of the BBS process is for team leaders, supervisors, managers, and executives to increase behaviors that will support BBS and improve safety in general.

It is important to understand the difference between direct behaviors that affect safety (what you do to keep yourself and your immediate peers safe) and indirect behaviors that affect safety (for example, management decisions about fixing equipment, delaying repairs, training versus no training, or emphasizing productivity over safety.) BBS processes address all of these behaviors. Depending on your job, you will focus on either direct or indirect behaviors (in some cases both). For example, if you are a frontline worker you will work on direct safe behaviors such as wearing personal protective equipment, consistently using machine guards, and bending your knees when lifting. If you are a frontline supervisor or team leader, you may have a few direct safe behaviors if you do some frontline work, but you will certainly have indirect behaviors as well. These indirect behaviors might include discussing safety as often as you discuss productivity so that employees understand that safety is a priority, using positive reinforcement for your teams' use of safe behaviors, and providing the necessary time to conduct safety training. If you are a senior manager, you will likely work almost exclusively on indirect behaviors such as analyzing how all decisions you make will impact safety, allocating resources toward improving safety, and holding your managers accountable for their indirect safe behaviors.

The word *direct* should never be confused with the word *intentional*. Many direct at-risk behaviors are unintentional, a product of trying to get the job done faster or even the result of management-controlled consequences. Behavior-based safety is not a process founded on the basis of blaming employees. It is a process that emphasizes using data to determine root causes and changing behavior at whatever level is appropriate to ensure that employees do not get hurt.

In summary, the goal of BBS is to provide your organization with a systematic and positive method for increasing and sustaining the behaviors, both direct and indirect, that will prevent injuries and accidents.

Observations

If you have heard anything about BBS, you know that it involves observations of work as a way to collect data on how frequently groups of people are behaving safely. This observation component often makes people nervous. They wonder if it is a policing system that requires peers to tell on each other. While some BBS processes may do that, they shouldn't, and a BBS process that is truly based on the science of behavior *absolutely* does not do this. BBS is not about catching people doing unsafe things and reporting same to management; in fact, it is the opposite. BBS is about checking to see how good people already are, working with them to improve, and then recognizing and celebrating that improvement every step of the way. All data is group-oriented and observations are anonymous; that is, *no names* are recorded. This no name/no blame observation method is simply a measurement tool that provides information about how safe the group is, because without data we can't recognize and celebrate improvement. With BBS, the safety team members collect group scores and the group works together to improve on critical safe behaviors. The frontline groups, not management, handle observation data. Managers work on their own set of indirect behaviors for supporting the safety process. This process is not about Big Brother; it is about getting an accurate picture of how the group is doing, and then working together to improve and receive recognition and reward—in addition to the reward of returning home safely to your family—for your efforts and achievements.

You are undoubtedly wondering if BBS really works or if it is another pain-in-the-rear exercise that adds more work to your day. At first (but only for a few days) it may be both. Is BBS eventually worth the time, energy, and resources invested? Of course, you will only know once you try it. However, people in a variety of industries have discovered that the process, as described in this book, takes a minimal amount of time and adds fun to the workplace. In other words, safety has improved and so has the workplace in general because people are given a voice and an active role in their own well-being at work. You will read case studies throughout this book that show you the power and efficiency of BBS.

Summary

- The overarching goal of any safety process, including BBS, is to reduce or eliminate injuries. BBS achieves this goal by increasing the number of safe habits of all employees and ensuring important support, or indirect behaviors, on the part of management.

- Observations are used as a means of collecting data in BBS. These observations are anonymous and used to provide meaningful feedback and positive reinforcement for all employees.

- Hundreds of companies in a wide variety of industries have used this BBS process successfully.

Did You Know?

- American businesses spend billions of dollars annually as a result of workplace injury, funds that could have been used on employee pay raises, bonus plans, profit sharing and so on. That amount doesn't include the indirect and long-term costs of retraining, turnover, diminished or lost skills, human suffering, lower profits and sales, property damage and the impact of negative customer perceptions of an organization's product and its employee care.

- According to OSHA and the U.S. Department of Labor, the ratio of indirect to direct costs of on-the-job injury varies from 1:1 to 20:1 and the lower the direct costs of the injury, the higher the ratio of indirect to direct costs. For example, the average cost of a minor eye injury is $1,463 when indirect costs are considered.

- La Paz, Bolivia, which is about 12,000 feet above sea level, is nearly a fireproof city, and the fire engine ordered out of civic pride gathers dust in their firehouses. At that altitude, the amount of oxygen in the atmosphere barely supports fire.

PART II

WHAT WE KNOW ABOUT BEHAVIOR

CHAPTER 3

RISKY BEHAVIOR: WHY DO PEOPLE BEHAVE IN AT-RISK WAYS?

"On my way home from work I saw some guys working in a trench and they didn't have any of the required safety protections. I thought, 'If that thing collapsed . . . .' Probably they haven't been hurt before and they don't think it's going to happen. And many times doing a job the safest way is actually punishing. But to have someone go through a near-death experience in order to become a safety champion is not something that we want to do. We have to recognize that doing a job safely may take a little longer or may cost a little more."

The above scenario, described by Kevin Sprague, an HSE manager at a polymers plant in South Carolina, probably occurs at work sites across America every day, but unfortunately, and too frequently, the only time such scenes receive attention is on the evening news—following a tragedy.

The irony is that many companies have complete and detailed rules and regulations concerning safety for every work site and every type of performance, but people continue to be hurt on the job. Why, after all of the painstaking attention to developing safety standards, is this still the case? The first step in finding the answer to this question is to understand that rules and regulations, after all, are but a set of instructions—in behavioral terms, these instructions are known as *antecedents*.

If you want to understand why people do at-risk (or unsafe) behaviors, or in fact if you want to understand why people do any and all the things they do, you need a basic understanding of the science of behavior (Applied Behavior Analysis). Behavior analysts study why people do what they do. Undoubtedly you have wondered why some people behave in certain ways and you have probably even wondered why you do some of the things you do. Well, the answer lies in something called the ABC or Antecedent, Behavior, Consequence Model.

THE ABC MODEL

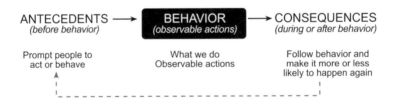

ANTECEDENTS → BEHAVIOR → CONSEQUENCES
(before behavior) *(observable actions)* *(during or after behavior)*

Prompt people to act or behave

What we do Observable actions

Follow behavior and make it more or less likely to happen again

- An *antecedent* is a person, place, thing, or event coming before a behavior that sets the stage for you to perform that behavior.

- A *behavior* is what you see when you observe someone working or what a person does.

- A *consequence* is an event that follows behavior or happens simultaneously with behavior that changes the probability the behavior will occur again and/or will affect the frequency or rate of that behavior.

Antecedents

Historically, companies have used a number of antecedents in an attempt to improve safety: putting up signs, writing procedures, requiring training, holding safety meetings, and so on. These are all antecedents. They prompt behavior. But do they result in permanent behavior change? In your experience, if your company introduces a new procedure for some work process, does that mean everyone automatically follows the new procedure from that day forward, 100 percent of the time? What about if you have a safety meeting about bending and lifting properly? After the meeting, does everyone start lifting properly and continue lifting properly forever? Most people agree that antecedents have a short-term impact, at best. What happens is that most of us follow the new procedure (lifting properly) for a while—maybe for a few days or even a few weeks. Unfortunately, even if we want to make the change, even if we truly believe that we should follow that procedure, it is difficult to change established habits. Without even being aware of it, we slip back into our old ways.

It's too bad that antecedents alone aren't enough to change behavior permanently. It would be so much easier if all we had to do was put up a sign, have a meeting, or hold a training class. Unfortunately those things alone are not enough. That doesn't mean that antecedents are bad. In fact, antecedents are necessary. We need training, meetings, instructions, procedures, and so on. However, if that is all we do, then we aren't going to get the kind of permanent behavior change we need to keep people safe. It may be okay in other parts of life to do the right thing most of the time or even some of the time, but in safety that isn't enough. We must make sure people engage in safe behaviors every day, all day. This is the only way we can be sure they won't get hurt.

Returning to the ABC Model, there are two things that influence behavior: antecedents and consequences. Since antecedents have only a temporary effect on behavior, we must turn to consequences if we want to achieve permanent behavior change. To do this we must determine what happens to performers when they follow a new procedure. Does it take more time? Is it a hassle? Does it make the work more difficult or easier? These are all consequences and they will have a bigger impact on what performers do and continue to do than any antecedent ever will.

All Consequences Are Not Created Equally

The ABC Model makes analyzing behavior sound easy, but it is not quite as easy as it seems. Every behavior, even a simple behavior, has multiple consequences. For example, the behavior of eating a donut results in multiple possible consequences. Some of those consequences are positive and some are negative.

CONSEQUENCES OF EATING A DONUT

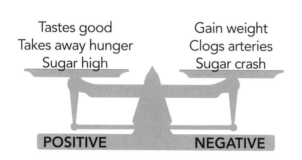

Tastes good
Takes away hunger
Sugar high

Gain weight
Clogs arteries
Sugar crash

POSITIVE   NEGATIVE

When you read the consequences listed above, you may be thinking that they are not all meaningful consequences for you. Of course, there may be more or different consequences depending on whose behavior we are analyzing. For example, a diabetic may suffer much more serious consequences from eating sugar than a non-diabetic. One of the things that makes understanding and changing behavior challenging is that we are all different. What is positive to one person may be negative to someone else. You may not be concerned about clogging your arteries, for example. Let's assume for now that the consequences listed are the right ones for someone named Joe. You'll notice there are three positives and three negatives. In other words, there are three consequences that would encourage Joe to eat a donut and three consequences that would discourage Joe. How will Joe ever decide what to do?

Well, not all of the consequences are equal are they? Some are more powerful than others. What is the difference between the taste of a donut and gaining weight from eating a donut? If you said that one is immediate and one is not, you would be right. The donut tastes good right now but you probably won't notice the weight gain until later. Here's another difference: the good taste of the donut is pretty certain (99.9 percent of all donuts we have eaten taste good), whereas gaining weight from eating one donut is uncertain; that is, it won't happen for sure. You might eat a donut, but skip lunch or go for a run later and burn off the calories. With those two differences (and your own experience), which consequence is more powerful: the taste of the donut or the potential weight gain from a donut? For most of us, the taste wins out every time.

Consequence Power

To help us see the potential impact or power any consequence will have on influencing behavior, we can categorize consequences in terms of the following characteristics.

TYPE OF CONSEQUENCE:

p OSITIVE to performer

n EGATIVE to performer

TIMING OF CONSEQUENCE:

i MMEDIATELY while the behavior is happening or immediately after

f UTURE a few hours, days, or later

PROBABILITY OF CONSEQUENCE OCCURRING:

C ERTAIN the consequence will happen close to or at 100 percent

U NCERTAIN the consequence only happens some of the time

Positive vs. Negative

Whether a consequence is positive or negative varies from one individual to the next. We are all different in terms of whether particular consequences will have a positive or negative impact on us. For example, some people love public praise and will work to get *more* of it. Other people are embarrassed by public praise and will work hard to *avoid* it.

A positive consequence is one that encourages more of the same behavior and a negative consequence is one that discourages more of the same behavior.

Immediate vs. Future

Consequences that are *immediate* are much more powerful than those that are *future*. The further away a consequence occurs in time following a behavior, the weaker its influence on behavior. This is one of the reasons that the threatened consequence of lung cancer and/or wrinkles is not very effective at persuading teenagers to stop smoking. Teenagers are relatively certain they will not get lung cancer the next time they light up a cigarette. They may get cancer in 40 or 50 years, if they get it at all. And no teenager imagines that he or she will ever have wrinkles! On the other hand, the immediate consequences for smoking (from the teenager's point of view) include such payoffs as looking cool, fitting in, even reducing one's appetite and staying thin.

Certain vs. Uncertain

Consequences that are *certain* are much more powerful

than those that are *uncertain*. A good example of a certain consequence is burning your hand if you touch a hot stove. This will happen every time you touch a hot stove. Anyone with children knows it usually takes only one experience for a child to learn not to touch the stove again. In this case, a painful injury proves to be a powerful consequence because it is certain and immediate. A good example of an uncertain consequence is getting a speeding ticket. You could speed day after day for years and not get a speeding ticket. Getting pulled over and receiving a ticket are uncertain. Anyone who drives in most U.S. cities knows that speeding tickets aren't effective at getting everyone to drive the speed limit.

PICs and NICs

Putting all the pieces together, we can analyze the power or strength of any consequence by determining whether it is positive or negative, immediate or future, certain or uncertain. As the Relative Power of Consequences diagram shows, consequences that are both immediate and certain (regardless of whether they are positive or negative) are the most powerful. Consequences that are positive, immediate but uncertain are the next most powerful. Think of winning at the slot machine or catching a fish. Both of those are P, I, but U and yet they are very powerful consequences. Consequences that are negative, immediate, but uncertain are less powerful. Think of using intermittent punishment with your kids. It's less effective, right? Those consequences that are certain but future (whether positive or negative) are also less powerful. Consequences that are both future and uncertain are the weakest of all.

THE RELATIVE POWER OF CONSEQUENCES

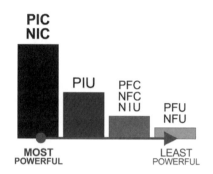

Summary

- The ABC Model helps us analyze consequences in order to understand why people do what they do including doing safe and at-risk behaviors.

- Antecedents alone result in temporary behavior change at best. Consequences must be used if permanent behavior change is desired.

- Typically, multiple consequences exist for every behavior, but not all consequences are equal.

- Immediate and certain consequences have more impact on our behavior than future and uncertain ones. Positive Immediate Certain (PIC) consequences and Negative Immediate Certain (NIC) consequences are the most powerful while Positive Future Uncertain (PFU) and Negative Future Uncertain (NFU) consequences are the weakest.

Did You Know?

- There is a town in Maryland called Accident. It sits on a major state highway linking western Maryland with the rest of the state. On the approach to the town is a road sign that doubles as a warning and is always true, no matter what the traffic condition. The sign says: ACCIDENT AHEAD.

- On an average day, Americans spend $2,021,918 on home exercise equipment; $3,673,973 on vitamins; and $10,410,959 on potato chips.

- One traffic accident every 10 years is what the typical U.S. driver can expect if the law of averages applies.

- The Anti-Careless Accident Campaign of Southwestern Australia sponsored a marathon in which a crew of young doctors and nurses were to push a hospital bed from Perth to Hobart. A nurse fell in front of the moving bed and broke her neck.

From Richard Hawk and Company, "Safety Stuff" Newsletter. Reprinted by permission of the author. www.makesafetyfun.com

CASE STUDY

Safety to the Forefront

Kevin Sprague, who remarked on the unsafe behavior of a road crew at the beginning of this chapter, soon came to understand that unsafe acts are supported by the immediate physical and cultural consequences in the work environment. "Behavior-based safety (BBS) is really changing our culture. People talk now and we're getting safety issues out in the open," he said.

Sprague, the health, safety, and environmental manager for a polymers plant in Greenville, South Carolina, works in a plant that manufactures carbon fibers and engineered resins. He describes the facility as a cross between a chemical plant and a manufacturing plant filled with movement, operator/equipment interaction, and the potential for injury that comes with the territory. His company also operates a plant in Rock Hill, South Carolina. The corporate goal to incorporate a BBS element into the safety process at all facilities led to a BBS pilot at the Rock Hill facility about two years ago. With BBS, employees identify safe behaviors, observe and collect data on those behaviors, and, with ongoing feedback, positive recognition, and reward, make those behaviors a habit among the workforce, defined for each team at the polymers plant as 30 scheduled shift days with 100 percent safe behaviors observed.

"What really caught our attention was that within the first year at Rock Hill we eliminated injuries. We simply didn't have anyone getting hurt," said Sprague.

"We also found that people seemed to like BBS. It was very positive, got people focused, and we were able to really use it to help the employees be safer."

A Telling Assessment

Before starting the behavior-based safety process, the Greenville plant conducted a baseline safety culture assessment, surveying employee perceptions regarding the safety process. One of the questions asked was, if you see someone doing something unsafe, do you say something to them about the unsafe behavior?

Almost every employee surveyed answered yes, but no one could remember a specific example of ever doing so or of receiving feedback for acting unsafely. People also answered universally in the affirmative when asked if they positively recognized others for acting safely, but no one could remember being positively recognized for acting safely. The results of the assessment showed that what people often think they do is not actually the case, a fact that reinforced the need for measurement and data collection.

Driving the Process

Plant management wanted the safety process to remain employee driven; but, for the process to have maximum impact, every employee must be actively involved. One successful innovation of the Behaviors Committee (developed to drive 100-percent employee participation) was to tie the safety portion of the hourly bonus to the BBS process. The Behaviors Committee developed a

BBS process checklist for monitoring employee participation. The committee designed the checklist so
employees could be caught doing the process right and
then get rewarded for it. The employees used the checklist and involvement in the process increased.

Another form of recognition—the BBS Hall of Fame—
continues to be popular. Published on the company's
internal Web site, the Hall of Fame lists the name of
every team that achieves a safe habit goal, the name of
the team's BBS champion, the goal the team achieved,
and details on what the group did to celebrate. The company also allocates a set amount of dollars per team
member to celebrate reaching safety subgoals and goals.
It isn't a great deal of money, but enough to enjoy a
lunch together, compliments of the company.

Much of the reward, however, comes from the social
enjoyment and the fun of achieving goals together. For
example, one unit at the plant designated the behavior
of walking through the pedestrian doors rather than the
roll-up doors designed for machine traffic. Anyone who
forgets and uses the wrong door can expect a round of
good-natured kidding from the team. "It becomes sort of
a game and that makes it fun," Sprague commented.

A Good Problem to Have

The problem at the Greenville plant is a good one to
have. With nearly 400 safe behaviors at habit level, the
people are running out of workplace behaviors to target
and are taking the safety process to the streets.
Wearing safety belts on the commute to work, following

at a safe distance in traffic, looking right and left before proceeding, and other specific behaviors, called *pinpoints*, are now in progress.

When Sprague spoke at the American Society of Safety Engineers (ASSE) conference, he discovered that many attendees were interested in a behavior-based safety process, but found that many such processes were too complicated. "People came up to me and said that the BBS process we used sounded much easier compared to others and that it made sense," he said. "I think the beauty of this process is that it can be tailor-made to coexist with other safety initiatives."

The Greenville plant's leadership was confident enough to examine existing policies and to then change those that were counterproductive to safety. The safety initiatives already in place were strong ones, addressing safe work conditions, but negative reinforcement also came into play and people in the plant didn't really own the process. Also, the plant leadership realized that the best of environments doesn't protect against unsafe acts or unsafe repetitive behaviors.

"I think the real key in the process is one of involvement by everyone," he said. "Another thing that is very important is that the leadership team has to have the guts to change, because we were doing a lot of things that were creating the wrong behaviors in people. It takes guts to admit that you could be wrong.

"Management walks a fine line between getting the process going and turning the employees into safety leaders. That takes time and patience. But if we are serious about eliminating pain and suffering, we need to give the tools to all employees so they can be successful in helping themselves and their teammates change unsafe behaviors to safe behaviors."

CHAPTER 4

CATEGORIZING CONSEQUENCES

Using the categorizations we learned in Chapter 3, we can complete a PIC/NIC Analysis®. The PIC/NIC Analysis helps us understand why people are currently doing what they are doing and helps us see what to do if we want to help people change that behavior.

This analysis is particularly important in safety where it is sometimes very difficult to understand why people do at-risk things. In most cases, people do an at-risk behavior because there are **PICs** for that behavior—Positive, Immediate, Certain consequences that encourage the performer to do the behavior. These consequences are often things such as the behavior gets the job done faster (PIC), is more comfortable (PIC), and/or easier (PIC), or managers (usually unintentionally) reinforce risk taking. There can be negative consequences but they are sometimes future (such as carpal tunnel syndrome or long-term muscle damage) and such consequences are almost always uncertain.

Even though most accidents happen while you are doing (immediate) the at-risk behavior, it is also highly uncertain that they will occur. For example, having an object fly up and hit you in the eye (N) while you are mowing the lawn without wearing safety glasses is a fairly immediate (I) consequence to the behavior of starting the lawnmower without the glasses but is highly unlikely to happen (U). So, it is easier (PIC) and more comfortable (PIC) to mow the lawn without glasses and the probability of getting an eye injury is very low (NIU). That is why many of us mow the lawn without wearing protective glasses.

On the flip side, most safe behaviors have NICs associated with them. That is, most safe behaviors have negative immediate certain consequences that discourage the behavior. For example, most personal protective equipment is uncomfortable (NIC) and inconvenient (NIC) in some way. Gloves reduce dexterity and make it harder to work (NIC); glasses can obscure vision (NIC); protective clothing can be hot (NIC).

> **We have just described the safety dilemma:**
> Safe behaviors often have negative consequences, and at-risk behaviors often have positive consequences.

This is exactly the opposite of what we want. If it were naturally reinforcing to do things safely, if there were lots of PICs for safe behaviors, then everyone would behave safely. Unfortunately, the opposite is more often the case. Most safe behavior can be naturally punishing in that many

NICs exist for safe behavior. This is a big reason why we still have at-risk behavior happening in the work world today despite an abundance of education, rules, and regulations. People know what to do and why they should do it, but the day-to-day consequences discourage them.

Just Trying to Get the Job Done

If you observe the at-risk behaviors that occur in your workplace, you will discover PICs for those behaviors. By and large, people do not do at-risk behaviors because they want to get hurt or because they are stupid. People do at-risk behaviors because positive consequences exist for doing so. In fact, people often perform behaviors at work to be better performers. They will get a job done faster or get more work done if they do things the unsafe way. Granted, this may only be perception. Possibly, it really isn't slower to do the tasks the safe way, but people often make decisions based on the perceived consequences for their behavior.

The fact that employees often take safety shortcuts to get work done more quickly is an important point. When this is the case, we should look at why getting work done faster is reinforcing. This is often a result of management behavior. If a manager or supervisor consistently recognizes employees who get more work done or who get work done faster, then those consequences are partly to blame for the worker taking shortcuts and not following all safety procedures. Can you begin to see how important the role of supervisors and managers is? While they are not on the floor telling employees to do unsafe behaviors, the contingencies they set up around production may unintentionally have that very effect. To change such behavior, managers and supervisors must provide consequences for behaving safely with the same frequency and intensity as those consequences they provide for production. Fortunately, BBS provides them with plenty to reinforce since safe behaviors happen many times every day.

Uncovering Barriers

Doing a PIC/NIC Analysis is an important exercise because it helps us understand why people do at-risk behaviors. If you do not understand why people do an at-risk behavior, then you cannot design the best method to make the needed change. This is why organizations often try to fix safety problems with training. They see, for example, that workers are not using a harness when working at a high elevation. Without doing an analysis, managers assume people aren't using the harnesses because they don't know they should or have forgotten. Thus, the company institutes a training or retraining program.

Occasionally, the problem may be training, but usually that's not the case. A PIC/NIC Analysis would probably reveal that getting a harness and putting it on properly are behaviors that result in NICs. It takes time (NIC); it is uncomfortable (NIC); it is difficult to find a place to tie-off (secure the harness to a second stable point) (NIC); and, being tied-off restricts movement and makes it harder to work (NIC). If we understand that the consequences are the real problem, then we can design a strategy that is more likely to be successful.

In fact, a PIC/NIC Analysis could also be called a Barrier Analysis because it uncovers the barriers or obstacles that prevent performers from doing the safe behavior. While the word *barrier* often refers to physical barriers (such as equipment in need of repair, poor lighting, lack of PPE available), a barrier is anything that stands in the way of safe behavior. If a safe behavior is overly time-consuming, uncomfortable, difficult, or a hassle, those are barriers to doing that safe behavior. In other words, barriers are NICs. Removing barriers (NICs) should always be the first strategy for changing behavior.

Often a PIC/NIC Analysis uncovers barriers to behaving safely that can be taken care of rather quickly. In one company, employees didn't want to take the time to find a ladder when certain repairs were necessary. Instead they climbed up the most convenient piece of equipment to do repair work, risking injury. A PIC/NIC Analysis revealed that getting a ladder involved several NICs. The ladders were kept in a locked room fairly far away from where the work was done. In addition, there were only a few ladders; often, even when people went to get a ladder, there were none available. Clearly these are the types of barriers (NICs) that can be dealt with by management. In this case, the organization purchased more ladders and placed them in several locations close to where the work was done. Problem solved! While not every PIC/NIC Analysis results in such simple solutions, doing the analysis will usually point you toward the best solutions.

Focus on Immediate and Certain

Knowing what you know now about the impact of immediacy and certainty, how effective would an annual safety banquet be at getting people to bend and lift properly every day? How about a five-year award for safe driving? While these events may be good for other reasons (improving morale and general recognition of working safely), they are not very effective at motivating people each day to make safe choices. A truck driver starting his first year with a company is unlikely to get in his truck each morning and say to himself, "I'm going to drive the speed limit today so that I can get that five-year safe driving award!" If we want to encourage people to do specific safe behaviors each and every day, then we should address the consequences that occur each and every day. We should reduce the NICs and increase the PICs for safe behaviors.

Adjusting the immediate and certain consequences for safe behaviors is exactly what BBS is designed to do. You may have wondered why it is important in BBS to do observations every day or why it is best to update a feedback graph every day. (More detail about feedback will follow.) Each BBS element is carefully designed to enable the delivery of more immediate and more certain consequences for safety. If there are fewer NICs (barriers) for safe behavior and more PICs, people will do those safe behaviors more often and therefore reduce the chances of anyone getting hurt. Again, the ultimate goal is to get those behaviors happening 100 percent of the time, thereby eliminating the risk completely.

Summary

- The safety dilemma shows us that there are often PICs for at-risk behavior and NICs for safe behavior.

- Safe behavior is often uncomfortable, more time consuming, or a hassle in some way.

- Barriers to safe behavior are often NICs.

- Doing a PIC/NIC Analysis allows us to see the consequences for safe and at-risk behavior so that we can design more effective strategies for change.

Did You Know?

- The first person to be killed in an auto accident in the United States was Henry H. Bliss, a 68-year-old real estate broker. On September 14, 1899, in New York City, Mr. Bliss stepped from a streetcar, turned to assist a woman passenger, and was hit by a cab.

- Benjamin Franklin's kite-flying experiment was a success. Not so for the next person who attempted the feat. Trying to repeat the sentry-box experiment, Russian physicist G. W. Richman was killed in St. Petersburg in 1753 when a "pale blue ball of fire, as big as a fist, came out of the rod" and struck him in the head. Richman died instantly from the lightning bolt and became the first martyr to the new age of electricity. (From: They All Laughed, by Ira Flatow)

- A small riding mower generates three times the throwing power generated by a .357 Magnum pistol.

From Richard Hawk and Company, "Safety Stuff" Newsletter. Reprinted by permission of the author. www.makesafetyfun.com

CHAPTER 5

THE CONSEQUENCES OF SAFETY

So far in our exploration of behavioral influences, we have discussed the limited effect of antecedents when we use them alone, established that consequences have the greatest impact on our behavior, and that immediate and certain consequences are the most powerful. In this chapter we will review the different types of consequences and the effects and side effects of each.

Generally speaking, consequences affect behavior in one of two ways. Some consequences *strengthen* behavior and some *weaken* behavior.

Consequences That Strengthen Behavior

The two consequences that strengthen behavior are both called *reinforcement*. *Positive reinforcement* is a consequence following behavior that strengthens that behavior. Examples of possible positive reinforcers include a smile, nod of the head, a positive comment, a T-shirt, a sense of accomplishment, or a cash award. Positive reinforcers, like all consequences, are defined not by the event or item (the reinforcer) but rather by the effect they have on behaviors. Anything that follows a behavior and leads to an increase in that behavior is a positive reinforcer.

Negative reinforcement is when we engage in behavior to escape from or avoid something. For example, if you are speeding on a freeway and see a police car, you are likely to take your foot off the accelerator and perhaps put your foot on the brake. You engage in these behaviors to avoid getting a ticket. *Negative reinforcement* is operating whenever a person engages in behaviors to avoid or escape something. This can be confusing, but remember that negative reinforcement also *strengthens* behavior in that we do more of the behavior that helps us avoid something bad. For example, procrastination (and the many behavioral forms it takes) is a form of negative reinforcement. One of the authors of this book gets a lot of work done around her house during income tax time. To avoid working on her taxes, she convinces herself that the house must be cleaned, the lawn must be mowed, the garage must be cleaned, etc. All of those behaviors are negatively reinforced because they result in avoidance of doing taxes, usually an unpleasant task.

The difference between negative and positive reinforcement is the *extent* to which they strengthen behavior. With positive reinforcement, a person will do more and more of the behavior to receive the desired consequence.

With negative reinforcement a person will only do the minimum amount of behavior required to avoid the undesired consequence.

Consequences That Weaken Behavior

The two consequences that weaken behavior are *punishment* and *penalty*. Punishment occurs when we receive something undesirable after a particular behavior and therefore do less of that behavior in the future or stop doing the behavior entirely. Getting injured or being reprimanded are potential examples of punishment. If an individual performs an at-risk behavior and gets hurt, he or she is less likely to repeat that at-risk behavior in the future. Punishment, like all consequences, is defined by the effect it has on behavior. Thus, if the individual in our example gets hurt but continues to do the at-risk behavior at the same frequency, then getting hurt was not punishing (or punishing enough) in that instance.

Penalty involves losing something that you want. For example, if you file your taxes late and have to pay a hefty fine, then you are more likely to file on time next year. In effect, the consequence that changed your behavior was a consequence of penalty. You lost something you wanted (your money) and you don't repeat the behavior that caused that loss.

Some events include both punishment and penalty. For example, getting a speeding ticket involves some punishment (get chewed out by police officer) and penalty (lose money when you pay the speeding ticket, possibly even lose your license). Similarly, having an accident at work may involve punishment (the pain of breaking your arm) and penalty (losing use of that arm for six weeks as it heals). Punishment and penalty are very similar in terms of the effect they have of decreasing behavior and in terms of the side effects they can create. Therefore, for the sake of clarity, we will refer to both under the singular category of *punishment* for the remainder of the book.

No discussion of consequences is complete without a discussion of *extinction*. Technically speaking, extinction is the absence of a consequence . . . the absence of any kind of reinforcement. Extinction occurs when the reinforcement we have been receiving or are expecting to receive doesn't happen. For example, you take extra time to ensure that the products you are shipping are packaged very carefully. After a while, you notice no one seems to notice your extra effort. You never hear anything positive from your team leader, your peers, or your customers. Gradually over time you are less careful about how you pack the products. This is extinction at work. A common thought we have when one of our behaviors is being extinguished is, "Why bother?" If we think no one notices or cares about what we are doing, then we may start to assume it doesn't matter, so why bother doing it? This is especially true when the behavior takes extra time and effort. We all have too much to do each day to continue doing things that don't seem to matter.

Extinguishing Good Behavior

Notice that with extinction it is irrelevant whether the behavior really does matter or not. The positive reinforcement and feedback we receive are the gauges we use at work (and in life) to determine what is important and what is not. This is true even if positive reinforcement and feedback (or the lack of either) conflicts with what we are told. In our example, the team leader may have told you that packaging the product very carefully is important. However, if that same team leader never says anything to you whether you package the product carefully or not, you begin to believe it really isn't that important. Furthermore, if the same team leader gives you feedback and positive reinforcement for high productivity (packing more boxes per hour), you will begin to spend time doing what produces reinforcement, possibly to the detriment of another task, such as packing the product carefully.

Extinction is all too common in our work lives. Many employees do a lot of good work every day that is rarely reinforced. This is especially true in the realm of safety. Unfortunately, safety has historically been managed reactively. In other words, companies sometimes don't pay much attention to safety when everything is going well. It's not uncommon for employees to relate that they only hear about safety when they have done something wrong. Yet, when they work safely they don't hear a word of praise or see a sign of recognition. Sound familiar? This circumstance can extinguish safe behaviors, allowing at-risk behaviors to take their place. As previously stated, people don't change safe behaviors into at-risk behaviors because they are lazy or because they want to get hurt. At-risk behaviors take over if the safe behaviors don't get reinforced (extinction) and the at-risk behaviors are reinforced because they are faster, easier, and more comfortable. This is an area where management can make a difference by ensuring that safe behaviors are reinforced and by minimizing the reinforcement for at-risk behaviors. *Important message: We must reinforce safe behaviors!*

Does this mean every safe behavior that every employee performs must be reinforced every time it happens? The answer is no. In fact, this is the reason the BBS approach focuses on developing *habits*. When a behavior becomes a habit, it is very resistant to extinction, so the goal of this process is to create habits that eventually require very little reinforcement. However, the paradox is that getting a behavior to the level of habit requires a great deal of reinforcement. We have to watch most diligently for extinction when a behavior is new. If we don't make sure there is enough reinforcement for the new safe behavior, that behavior may quickly disappear. Once the behavior occurs consistently, it is a habit and requires much less reinforcement.

Now that we have defined the types of consequences as positive reinforcement, negative reinforcement, punishment/penalty (to be discussed as punishment), we can explore each in more detail. To effectively implement behavior-based safety (or any other behavior change), you should understand each type of consequence, its effect on behavior, and its side effects. Unfortunately, when people don't understand consequences well, they misuse them, often resulting in undesired outcomes. As you will see, it is all too easy to elicit the exact opposite of your intended effect on behavior. Thus, a thorough understanding of consequences and the ability to recognize their influence at work (even at home) is critical.

Summary

- Two consequences that increase behavior are positive and negative reinforcement.

- With positive reinforcement, we do more of a behavior because something desirable happens to us.

- With negative reinforcement, we do more of a behavior because it helps us to avoid something undesirable.

- Punishment and penalty both decrease behavior.

- Punishment means we get something undesirable. Penalty means we lose something desirable.

- Extinction is the absence of reinforcement resulting in decreased behavior.

- It is important to guard against extinction when trying to develop safe habits by using plenty of positive reinforcement, especially initially when the behavior is not yet a habit.

Did You Know?

- Russian psychic E. Frenkel failed in his attempt to stop an oncoming freight train through the use of his psychic powers. The train engineer said that Frenkel jumped onto the tracks in front of the train with his arms raised and his head lowered. Investigators later found Frenkel's notes, which read, "First I stopped a bicycle, cars, and a streetcar. Now I'm going to stop a train." [This is a classic example of misinterpreting the behavior/consequence relationship!]

- An eighteenth century Parisian named Jean Jacques Perrett became tired of having his face cut while his barber shaved him. Wouldn't shaving be much safer and more comfortable, he wondered, if a wooden guard were attached to the straight razor blade so that only a snip of the blade protruded? And so was born the safety razor.

From Richard Hawk and Company, "Safety Stuff" Newsletter. Reprinted by permission of the author. www.makesafetyfun.com

CASE STUDY

Giving Safety a Lift
at the Grande Prairie Sawmill

Danger and *sawmill* seem to be naturally associated words, yet Weyerhaeuser Canada's Grande Prairie sawmill site holds one of the best safety records in the industry. However, being among the best in one industry isn't good enough, according to Doug Chappell, dry side manager. "We want to be the best in the world in all industries. So although we do benchmark ourselves against our own industry, we also benchmark ourselves against some of the world leaders in safety," he explained.

Weyerhaeuser Corporation is one of the largest integrated forest product companies in the world. Weyerhaeuser Canada, a subsidiary of Weyerhaeuser Corporation, operates 10 sawmills, three pulp mills, three oriented strand board (OSB) mills, and one paper plant. The Grande Prairie site runs an integrated pulp mill and sawmill.

Several years ago, the sawmill operation's leaders decided to include a safety breakthrough initiative as an integral part of the annual strategic business plan. The sawmill's 200 employees shared an admirable safety record, but Chappell and other sawmill leaders wanted to take safety to a higher level. "You can only do so much with engineering," he said. "You can have the safest plant in the world, but if people aren't behaving in a safe manner, they're still going to get hurt. In industry, when we start working on safety, we do all of the obvious things—engineering-type fixes, reengineering equip-ment, and creating a safer environment. Those things will drive your injury rates down to some level, but then you start to plateau. Until you get to the behavioral aspects of safety, you're simply not going to drive your injury rate below that plateau, and you certainly won't be working in terms of world-class results on safety."

At the Grande Prairie planer mill, two shifts of 20 people each produce up to 60,000 pieces of lumber per shift. The employees continually move lumber down a 300-foot single line in a process that requires repeated physical handling, stacking, flipping, and grading of each piece. One might call it backbreaking or, at least, back-straining labor, and five years ago, the data proved that description true. In a group of 40 employees over a 12-month period, seven suffered from varying degrees of back injury—the largest single type of injury in the facility.

The data strongly indicated that back injury should be a priority focus for the safety-initiative breakthrough process, but as Chappell stated, "As usual we did all kinds of work on the antecedent side." Soon employees were involved in the site's back power program that included proper lifting technique demonstrations, regu-lar back education videos, even an individual observa-tion/consultation (with a physiotherapist) for each employee. The education was valuable and the back injuries dwindled to zero, but Chappell had misgivings. "I felt like the absence of back injuries was a lot more good luck than good management. I wasn't as tuned-in to behavior as I am now, but I did notice that when I walked through the mill, I still observed people doing unsafe behaviors such as lifting things off of the floor the wrong way," he said.

The plant's leadership team attended several presentations on a variety of safety processes. They observed that many of the processes primarily targeted leadership activities—planning, future vision, leadership roles. "Behavior-based safety was the only one that really focused on how to change or manage behavior. The way I look at it is we can use some of the planning tools as 'Here's where we want to go,' but BBS is the way to make it happen," Chappell stated. The leaders of the Grande Prairie site saw BBS as a process that could be implemented quickly and that would give everyone a positive experience, thus aligning the workforce. Because the organization had relied on antecedent-heavy programs in the past, the decision was made to underplay the rollout but to educate everyone in the facility about behavioral management methods. "We wanted people to know there is a strategy behind this. We're not just playing games," Chappell explained.

We learned that the habit-level behavior in our plant was the unsafe one: bending at the waist to lift things off the floor," he said. "Back injuries usually result from long-term unsafe behaviors, so we wanted to replace the unsafe habit behavior with a safe habit behavior—proper lifting—and that's what we designed our process to do."

The Planer Mill Lifting Olympics

To try to let people know that the new safety initiative would be fun, not threatening, Chappell and other safety team members created the theme of the Planer Mill

Lifting Olympics. "I made a point of telling people that nothing bad would happen to them if they didn't participate, but good things would happen if they did," said Chappell. Baseline observations showed that, on a daily basis, only 28 percent of lifts were performed safely.

The group then established bronze, silver, and gold medal performance levels:

BRONZE: 100 percent participation plus 75 percent safe lifts for five consecutive days

SILVER: 100 percent participation plus 90 percent safe lifts for five consecutive days

GOLD: 100 percent participation plus 98 percent safe lifts for five consecutive days

After pinpointing the safe/at-risk lifting behaviors, everyone received a set of observation cards, illustrated with an Olympic torch and a weight lifter, designed to be easily completed within 10 seconds. A large poster blocked off the days and listed each mill worker's name. To receive a sticker credit, each person needed to complete a minimum of one observation per day. Within a few days every employee became involved in the safe lifting observations. But Chappell didn't allow that fact to let him or other leaders off the hook. Each manager set a goal to recognize and talk to a minimum of four individuals per day regarding their involvement with the process. "We especially targeted people for recognition who had not participated one day but did participate the next," Chappell stated. "In other words, we tried to get as close to immediate reinforcement as possible for the new behavior of participation."

The observation cards included check boxes for whether a safe or at-risk lift had been observed and a yes or no box for whether the observer provided feedback to the performer. The feedback data helped determine how much feedback (if any) people were receiving out on the floor. The observer then signed his/her name, not the name of the performer, and turned in the card. Chappell was amazed that the 100 percent participation happened within three days, a rate that he thought impressive for a crew of 40 people. He also remained cognizant of indicators of performance that occurs under positive or negative reinforcement conditions. The minimum requirement for meeting the 100 percent participation criteria for the group was 40 observations per day, but Chappell never received less than 45 cards and often as many as 75 per day. "That told me that people were participating because they wanted to. They were doing more than the minimum and that was reinforcing for myself and others who were trying to make this work," he explained.

With daily group feedback and fun forms of encouragement, reinforcement, and recognition, the lifting Olympics pressed quickly ahead. The bronze medal celebration occurred within two weeks of beginning the initiative—indicating a jump from 28 percent safe lifts to 75 percent. The Olympic Lifting emphasis began in June. By October the mill had sustained over 98 percent of safe lifts for seven weeks. Chappell believes that a majority of the mill workers developed habit-strength safe lifting habits as a result.

He is well aware that self-report could involve some cheating, but he dismisses that as an irrelevant factor at the mill. "The bottom line is that if people want to cheat, they will," he said. "When we rolled this out, we told everyone we knew it was very easy for them to simply make up observation cards without doing the observation or to misinform us about the safety of the lifts. But we said, 'The reality is that we'd only be fooling ourselves. This is being done for you. This is a behavior we are trying to help you adjust to that will benefit you for a lifetime.'"

An admitted skeptic of former safety improvement results, Chappell now invites visitors to go out on the floor and watch the lifting behaviors of the mill employees. "I know the baseline and what the behaviors looked like before," he asserted. "I observe people lifting quite frequently. And this change is real."

CHAPTER 6

THE LOWDOWN ON PUNISHMENT

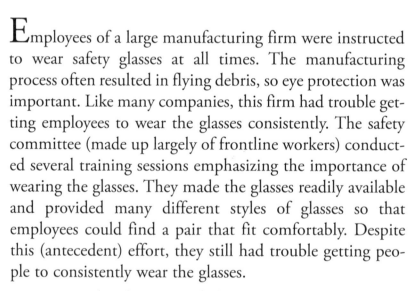

Employees of a large manufacturing firm were instructed to wear safety glasses at all times. The manufacturing process often resulted in flying debris, so eye protection was important. Like many companies, this firm had trouble getting employees to wear the glasses consistently. The safety committee (made up largely of frontline workers) conducted several training sessions emphasizing the importance of wearing the glasses. They made the glasses readily available and provided many different styles of glasses so that employees could find a pair that fit comfortably. Despite this (antecedent) effort, they still had trouble getting people to consistently wear the glasses.

Frustrated and concerned about their fellow employees' safety, the committee devised the following strategy. With the cooperation of management and the onsite safety professional, the committee members were given the authority to provide on-the-spot write-ups when they caught any employee working without safety glasses. These write-ups went directly into the employee's file. Armed with this new authority, the members of the safety committee began actively looking for people who were working without their glasses. The shop floor soon turned into a police state. The main goal of safety for peers deteriorated into the goal of catching someone who wasn't wearing protective eye gear. Productivity plummeted; resentment rose; but the wearing of safety glasses increased negligibly, yet only temporarily, and the entire safety process deteriorated into a contest between them and us.

> **This should not be the goal
> or the outcome of any safety process!**

The goal in behavior-based safety is to get people to behave safely more often, even all the time. Since most people acknowledge that they do at-risk things, at least occasionally, getting to the goal requires changing behavior. We know that consequences are the most effective way to change behavior and that we have several consequence options for doing so.

Once you understand all the consequences more thoroughly, you can better evaluate the probability of success of any strategy both short- and long-term. To review, we know that negative and positive reinforcement increase behavior but that negative reinforcement only increases behavior to the extent that the performer avoids a negative consequence or punishment. Positive reinforcement, however, maximizes behavior because the performer does the behavior at a high rate to receive the positive consequence. Punishment (or penalty) is anything following behavior that weakens or stops that behavior. For example, having someone criticize a suggestion you make may weaken your tendency to make suggestions. Also, following a long safety procedure may get you behind on your production quota, so you skip some steps in the procedure. These are common examples of punishment at work.

Punishment can be something quite significant such as a formal reprimand that goes in your personnel file, or it may be something quite small and subtle, like someone rolling their eyes when you speak up at a meeting. Remember that punishment is *anything* that decreases the behavior it follows, so it isn't just the big, serious stuff we tend to equate with the word *punishment*. In fact, many small punishers operate in the safety arena, such as the seemingly small, but immediate punishers made apparent through completing a PIC/NIC Analysis. Many safe behaviors receive punishment (and therefore don't continue to happen) either because they take time, are uncomfortable, are a hassle, or all of the above. Even though small, these consequences act effectively as punishers in that they decrease behavior. What makes these types of consequences (time consuming, uncomfortable, irritating) most effective, you will recall, is that they are *immediate* and *certain*.

Like all consequences, we can use punishment purposefully or inadvertently. We are using it purposefully when we take away our daughter's television privileges after she hits her brother. A supervisor is using it purposefully when he calls an employee into his office to discuss poor performance. Unfortunately, it is easy to use punishment unintentionally. An example is the manager who assigns more work to his hardest-working employee to make sure the work is completed on time and with high quality. High-level workers are often "rewarded" with more work, responsibility, and an implied promise of reward that never materializes. Is it any wonder that after awhile some of these hard workers learn not to work so hard?

Another example of unintended punishment: a team leader takes the time to have a safety meeting during the day, only to be informed by his manager that his team had a lower end-of-day production than other teams. The manager who did this probably didn't intend to punish the behavior of holding a safety meeting, but that is exactly what happened. As you may know, these types of situations take place all too often. That's why it is so important for everyone in an organization to understand consequences. If you don't even realize you are punishing a behavior that you want to occur, then how will you correct your actions?

In behavior-based safety, whether you are a manager, a supervisor, or a frontline employee, you must be careful not to punish safe behaviors or behaviors that promote safety. That makes sense to everyone, but what about at-risk behaviors? Since punishment decreases behavior, why not punish the at-risk behaviors? This is certainly a strategy used by many companies (either purposefully or inadvertently). As a general strategy, this is not a good idea for a number of reasons.

1. *Punishment tends to have a short-term effect on behavior unless it is immediate, certain, and rather severe.* Most of the punishment used at work does not meet these criteria.

2. *When you punish one at-risk behavior, you don't know which type of behavior will take its place.* It may be replaced by a safe behavior or it may be replaced by another, even worse, at-risk behavior.

3. *To use punishment effectively you must catch the performer in the act.* This means someone has to see the worker engage in the at-risk behavior. Supervisors and safety professionals know the problem with this, which is that under this type of regime, people may behave safely when they know someone is watching but behave unsafely the remainder of the time. Thus, effective punishment is a labor-intensive process. *If you never see the at-risk behavior, you can't use punishment.* This is an important point. If employees only behave safely when someone is watching, it means that some or most of the time they may not be behaving safely. If people aren't behaving safely all the time, then there is a chance someone could get hurt. Any strategy associated with safety must work all of the time, not just some of the time.

4. *Punishment is useless with lone workers because no one is available to "catch them in the act."*

5. *The overuse of punishment leads to some very negative side effects.* When people feel like the safe behaviors they do are never acknowledged but the at-risk behaviors are always noticed, any or all of the following may happen.

- Lower morale
- Lower productivity
- Decreased teamwork
- Lower trust
- Increased absenteeism
- Decreased volunteerism
- Increased turnover
- Desire to retaliate (sometimes acted upon in small or significant ways)
- At-risk behaviors go underground. (People do the right things only when someone is watching.)

When you combine these factors, punishment really isn't a viable option. If the goal is to encourage people to behave safely all the time, then punishment simply isn't a good choice.

That said, it *is* appropriate to use punishment in particular instances, but not as a general strategy. One such instance is when an employee is doing something that puts him/her or others in imminent danger. For example, if an employee is about to do something that will cause an explosion, it is wise to stop that behavior immediately. Punishment is appropriate. Another instance is when an employee is purposefully violating a safety rule. This is unacceptable behavior and failing to deal with it through immediate punishment (reprimand, suspension) risks condoning it. In addition, when supervisors or managers see at-risk behavior they are legally obligated to intervene.

Most people agree that punishment is the right consequence in cases like those listed above. But think about how often these types of things happen in your workplace. For most of us, these are rare events. That means that punishment should also be a rare event. Most of the time people do at-risk behaviors because they have developed an unsafe habit or they aren't aware they are doing it. Such behaviors can be dealt with most successfully by using positive strategies that lead to permanent behavior change. BBS focuses on unintentional unsafe habits, since these types of safety infractions represent the majority of at-risk behaviors at work. In summary, while punishment has a legitimate, though limited function, behavior-based safety emphasizes positive reinforcement and recognition rather than punishment.

Summary

- Punishment weakens behavior.
- Punishers can be big or small, purposeful or inadvertent.
- Using punishment in safety requires catching people in the act, which is difficult to do.
- Overuse of punishment leads to negative side effects.
- Punishment is appropriate under some circumstances, but the majority of the time behavior change can be accomplished without punishment and its side effects.

Did You Know?

- The famous conductor, Jean Baptiste Lully, had a habit of pounding out the rhythm to his orchestra by stamping a long, pointed cane on the floor. One morning, he missed the floor, speared his foot, contracted blood poisoning, and died.

- The world's first traffic lights were erected in 1868 near the houses of Parliament in London. They consisted of revolving green and red gaslights. They blew up soon after they were installed, badly injuring the policemen operating them.

- A drum major hurled his baton high into the air during a parade in Ventura, California. The baton hit a power cable, melted, blacked out ten blocks, put a radio station off the air, and started a grass fire.

- A man using an outhouse near Lawrence, Kansas, lost his balance while trying to retrieve his wallet, which had fallen through the floorboards. He fell in and had to spend seven hours in three feet of muck before being rescued. Douglas County Sheriff Loren Anderson described the man as unhurt, "but in a pretty ugly mood."

From Richard Hawk and Company, "Safety Stuff" Newsletter.
Reprinted by permission of the author. www.makesafetyfun.com

CHAPTER 7

NEGATIVE REINFORCEMENT: THE HAVE-TO SYNDROME

"I have to get to work on time or I'll be fired."

"I have to finish this report so the boss won't be mad."

"I have to rake these leaves so Dad won't yell."

"I have to wear these safety glasses
so the supervisor doesn't write me up."

As stated previously, negative reinforcement does work, because it makes us do things that we might not otherwise do. However, it isn't the best consequence option if we truly want to inspire the best performance. Why? Because negative reinforcement occurs when we engage in a behavior to escape from or avoid something bad. In other words, we do something because we feel like we have to *or else*. This means that we aren't going to go out of our way to extend our behavior past simple compliance. For example, if your manager makes derogatory remarks when you don't meet quota but doesn't say a word when you exceed quota, then just meet quota. Save yourself the extra effort!

And there's the key—extra effort. You'll never get extra effort using negative reinforcement.

Most of us set an alarm clock at night before we go to sleep. Do we do that because it makes us happy to hear that alarm at 6 a.m.? In most cases people set their alarms to avoid being late for work. This is negative reinforcement. You engage in the behavior (set the alarm) and you continue to engage in the behavior to avoid something bad (being late). Another example is putting oil in your car. Do you do this because it's fun? Well, a few people might find it fun which would make it positive reinforcement, but most of us do this only to avoid the bad things that would happen to our cars if we don't do it. We do the behavior (the behavior gets reinforced) only because it allows us to avoid something we don't want. (Of course, in this case, negative reinforcement is a perfectly suitable consequence because we only need a certain amount of oil in our car and it isn't a behavior that we are expected to engage in every day.)

At work, negative reinforcement is sometimes referred to as *motivation by fear*. Fear is certainly one of the predominant emotions we experience under negative reinforcement. In fact, people often talk about negative reinforcement this way. We may drive the speed limit despite being in a hurry because we're *afraid* of getting a ticket. We set our alarms because we are *afraid* we will be late if we don't. It makes sense, therefore, that if we are not afraid of the outcome of a behavior, then negative reinforcement doesn't work.

Threats and Warnings as Antecedents

Often people attempt to use negative reinforcement on themselves or others by delivering antecedents we would generally call *threats*. Those threats are usually some form of the following statement: "Do it, or else." The police tell us to drive the speed limit or else get a ticket. Our mechanic tells us to put oil in our car or else we will ruin our engines. Sometimes negative reinforcement is subtle and sometimes it is very obvious. In either case, we are engaging in behaviors to avoid the *or else*.

A great deal of negative reinforcement exists in our everyday lives both at home and at work. We engage in many behaviors to avoid negative consequences. For example, we go to the dentist to avoid developing tooth decay and we eat carrots instead of potato chips to avoid gaining weight. However, most of us don't always do these things as consistently as we should because the negative consequences are future or remote. But if we get a toothache, we're at the dentist pronto in an attempt to avoid more pain. And those carrots come out more frequently if there is a high school reunion around the corner.

As you may suspect, negative reinforcement is a big part of safety in that most safe behaviors are a means of avoiding injury or worse. For example, if you work in a bakery and part of your job is taking baked goods out of the oven, you most certainly use oven mitts (or some other protection) when you do this. You don't wear the mitts because something wonderful happens to you (positive reinforcement); you wear them to avoid the negative and immediate consequence of getting burned. There are similar examples in every workplace.

Interestingly, behaviors such as wearing the oven mitts occur without intervention. While there may be some initial training, no coaxing or warning or reminders are necessary. People simply wear the mitts. The safety person doesn't have to issue ongoing threats: "Wear the mitts or else you will burn your hands!" Furthermore, no one has to be positively reinforced for wearing the mitts. Why is it that all safe behaviors are not like this one? Why can't we just rely on negative reinforcement to work? Why don't people do all the safe behaviors required at work simply to avoid getting hurt?

If we do an analysis of our bakery example, we can see that getting hurt (burning our hands) is *immediate* and *certain*. Remember that immediate and certain consequences are the most powerful of all consequences. The reason we don't have to remind people to wear oven mitts or positively reinforce them for wearing oven mitts is that the natural consequence of getting hurt when not wearing them is immediate and certain. Many safe behaviors at home and at work fall into this same category. If we don't do the safe behavior we will get hurt—right now and for sure. We don't need to worry about these types of behaviors. The natural consequences ensure that people do the right thing. If only it were so easy for all safe behaviors . . . but it is not.

Most of the at-risk behaviors that happen in your workplace continue to occur because the consequence of getting hurt is uncertain and, in some cases, both future and uncertain. This is why just warning people (an attempt at negative reinforcement) doesn't work. We have learned to ignore the warning because it is unlikely to occur. Sure the supervisor can insert some punishment by reprimanding you, say for not wearing safety glasses, *if* he catches you, but the immediate and positive reinforcement of not wearing the safety glasses outweighs the uncertain consequences of being caught or being injured.

When we warn people about the possibility of getting hurt and it doesn't work, we often beef up the negative reinforcement by adding more threats. The police add the possibility of getting a speeding ticket to try to get us to drive the speed limit. The hope is that if people don't drive the speed limit to avoid getting hurt, then they will drive the speed limit to avoid a ticket. The reason speeding tickets came about was as an attempt to insert a more likely negative consequence for speeding—a fine—to try and prevent an uncertain consequence—an accident. Has this approach worked? No, because the probability of getting a ticket is also low—unless you're from out of town.

This strategy of threatening negative consequences in hopes that people will do the desired behavior to avoid those consequences (negative reinforcement) is an often-ingrained strategy of the traditional workplace. Safety professionals, safety committees, managers, and supervisors warn employees of how they could get hurt in hopes of getting them to engage in safe behaviors to avoid getting hurt. The problem is that many employees are no longer afraid of getting hurt. If an employee has done a particular at-risk behavior for ten or twenty years and has never been hurt, that employee is likely to think that he/she will never get hurt. Unfortunately, low probability (uncertain) consequences teach us a bad lesson. They teach us it is okay to do the at-risk behavior because we keep getting away with it.

When warning and threatening people about the possibility of getting hurt doesn't work, managers and supervisors may threaten that written reprimands will be given to those who do not behave safely. They may even threaten termination. Negative reinforcement is only effective if the negative consequence is fairly immediate and certain. Telling employees they will be fired if they do not use safety procedures is only effective if employees believe they will actually be fired. If people have witnessed someone getting caught not following the procedure and that individual did not get fired, then the threat becomes an empty and ineffective one.

You have probably noticed that negative reinforcement and punishment have a lot in common. They are close cousins and are almost always used together. Some examples follow.

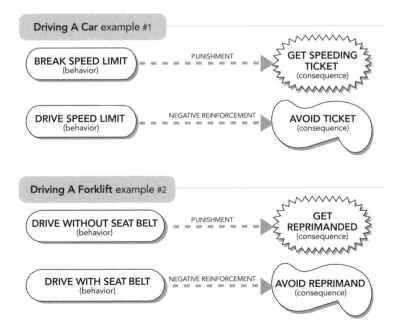

Often if we experience punishment for an at-risk behavior (we get hurt), we will change to the safe behavior to avoid getting hurt in the future. But we can't wait for everyone to actually suffer an injury to get people to change to safe behaviors. Negative reinforcement came into overuse because people hoped if we just warned workers about getting hurt, they would change their behavior(s). Experience shows that for negative reinforcement to work people have to believe that something bad really might happen to them. So, if the probability of getting hurt is low, the warning won't work.

Because negative reinforcement and punishment are often used together and are so closely related, they often result in the same side effects discussed in the punishment chapter: resentment, lack of trust, deteriorating productivity, retaliation, and lower morale. Thus, negative reinforcement should be used with caution.

When To Use Negative Reinforcement

Of course, as with all consequences there are appropriate uses of negative reinforcement. Negative reinforcement can be used when the behavior you are looking for (the safe behavior) never happens. If someone never puts on his or her gloves while working, then you have nothing to positively reinforce. Thus, you can use negative reinforcement to jump-start the behavior. Once the behavior occurs, even infrequently, you should switch to positive reinforcement.

While both negative reinforcement and positive reinforcement strengthen behavior, they have different long-term effects and different side effects. With positive reinforcement we get something we want as a result of our actions. With negative reinforcement the best thing that happens to us is we avoid something bad. Because of this, the effects of negative reinforcement and positive reinforcement are very different. With negative reinforcement people will perform only as much of the behavior as is necessary. We call this *just enough to get by* performance. We observe this firsthand when a police car is on the freeway and everyone drives the speed limit; but as soon as the police car pulls off the freeway, everyone speeds up again.

As with punishment, negative reinforcement requires that someone watch people at work all the time. Saying "do it or else" is ineffective if you are not around to see whether people act safely or not. Otherwise, how can one deliver the *or else?* So, as with its close cousin punishment, negative reinforcement is a labor-intensive management strategy, and it is not really an option for improving safe behavior at work. While there may be some circumstances when it is enough for people to do the bare minimum, that's not the case with safety. People should behave safely all day, every day, for themselves and for their families.

Summary

- Negative reinforcement is when we do things to avoid something bad.
- With negative reinforcement people do the behavior only when they have to, not all the time.
- Negative reinforcement and punishment are often used together, so result in the same negative side effects.
- The goal of BBS is for people to do safe behaviors all the time. Hence, negative reinforcement is not the consequence to use.

Did You Know?

- According to statistics, the best drivers are women from age 16 to 25.

- Compared to the population, emergency rooms treat nearly twice as many left-handed people for accidents as right-handed people.

- Each year 1 in 20 Americans will receive emergency room treatment for fall-related injuries.

- The added risk of fatalities when interstate rural speed limits in the U.S. went from 55 to 65 mph: 19 percent. Serious injuries increased by 40 percent.

CASE STUDY

BBS at Shell Western

Bob Lane, the general manager of Shell Western E & P Inc., recalled a near-miss situation on an offshore oil rig. "One of the crew was working with an autoclave fitting when it came apart and shot a very small but very high-pressure stream across a passageway," he said. "The worker happened to be standing in the right spot, but if he had been standing in the way, that stream would have shot a hole right through him and would probably have killed him."

This scenario stands out in Lane's mind because such incidents at his company are very rare. Many of its on- and off-shore oil and gas facilities hold multiple-decade records of zero lost-time accidents. However, Lane understands that it isn't wise to blindly trust such statistics. That crew member may have previously repeated the autoclave task 10,000 times or more while standing in a position that would have resulted in a different, and tragic, outcome.

"Statistics may say that chances of such a thing occurring are one in a million, but we want to take even that statistic out. We want people to stand in the right place when they're working on a fitting every time," said Lane.

Soon after the above occurrence Shell E & P began using behavior-based safety as a way to address potential problems upstream—a proactive as opposed to a traditionally reactive safety approach. The organization found that fear of an accident is difficult to sustain because accidents don't occur frequently enough for people to remain afraid of repeating at-risk behaviors. The heart of Shell's BBS process is refocusing people's attention away from fear of having an accident to the pride and satisfaction for acting safely, gained through the use of effective positive reinforcement and recognition.

Knowing what to positively reinforce, however, is critical. Shell's experience has revealed that it is dangerous to reward only positive safety results, such as the number of days without an incident. Such approaches promote (often unintentionally) negative consequences for reporting incidents, so that people hesitate to report minor injuries and near misses, giving management a skewed perspective on how safe the organization really is. Shell addressed this problem by announcing up front that reporting accidents would be viewed as a positive, preventive measure. To back up this assertion, management graphs the number of reported minor first aid incidents, shares feedback with employees, and reinforces their efforts to report all injuries.

At Shell, the ratio of lost-time OSHA recordables is regularly compared to and contrasted with total incidents (minor injuries). Since Shell began emphasizing the reporting of minor accidents as a critical upstream behavior, that ratio has more clearly reflected reality. With people confident to report minor incidents and with an emphasis on measuring and recognizing safe behaviors, Shell can use OSHA recordable rates as a checkpoint for whether they are working on the right behaviors rather than use those rates as the sole measure of safety.

"When you are watching people to make sure they bend or lift correctly, then you become aware of doing it correctly yourself," said Vic Craig, a maintenance aide and safety leader who works at a field location that has operated for decades without a lost-time accident. "It doesn't take five minutes out of a work day to stop and watch somebody do a job and mark a scorecard. We enjoy it too, because it's pinpointing the proper way of doing things instead of punishing the improper," he said.

Why such vigilance at a field with a record of no lost-time accidents in 30 years? "We want to nip it in the bud before it ever happens," Craig said.

PART III

HOW TO APPLY WHAT WE KNOW

CHAPTER 8

PINPOINTING:
SELECTING BEHAVIORS THAT COUNT

Now that you understand more about how to effectively influence behavior, how can you put this knowledge to use to improve safety? As we discussed in the first chapter, behavior is an integral part of safety. The right behaviors can keep people safe and the wrong behaviors can lead to injuries. In this chapter we will cover how to select behaviors to work on in BBS. We will start with safe behaviors that are most often performed by frontline workers and then move to behaviors that supervisors, managers, and executives should do to improve safety. Before we start, a few words about the BBS process. There are many different ways to implement BBS. Because the next few chapters focus on how to do the process, you will begin to see a bias toward Aubrey Daniels International's particular approach. While this approach has been extremely effective, it is certainly not the only approach. The good news is that the concepts and ideas discussed are relevant to most behavioral approaches.

Behavior-based safety is only as effective as the behaviors you select to measure, to provide feedback, and to reinforce. Since people do thousands of behaviors every day at work, it is impossible to address all of those behaviors (and not all of them need to be addressed). In fact, we advise that you work on only a few behaviors at a time. Once those behaviors become safe habits, then move on to others, revisiting the habit level behaviors occasionally for maintenance purposes. We find that three to five behaviors are a good amount. Why so few? Well, keep in mind your goal in BBS is to develop a safe *habit*. That often means breaking an unsafe habit. If you have ever tried to lose weight, start exercising, quit smoking, or stop biting your fingernails, you know that changing habits isn't easy. If someone asked you to change fifteen of your personal bad habits at once, you would feel overwhelmed, wouldn't you? Working on just a few behaviors at a time is more manageable both for the people trying to develop the safe habits and for the people doing the observations. It is quicker and easier to observe three to five behaviors than to monitor fifteen.

The first step is to decide where to begin. Obviously the best behaviors to select are those related to injuries and near misses that your work group has experienced. Sometimes groups are tempted to work on behaviors that are easy to observe but are not really behaviors related to any of the injuries their group has suffered. While this makes observation easy, it really undermines the whole process since there will be little reduction in injuries, and ultimately people will see BBS as a waste of time. It is far better to take the time to select behaviors that will really have an impact on keeping people safe and will be much more reinforcing in the long run.

BBS and Non-Repetitive Work

Some employees may think that because they are in a job with a multitude of duties, construction work for example, that BBS can't really apply. But of course, it can. In such work there are still important repetitive behaviors such as safe lifting, proper use of equipment, and wearing PPE. Furthermore, tasks such as securing the work area and doing quick hazard checks can all be improved using the BBS process.

Behavior Sources: Accident/Injury Rate and Near-Miss Reports

The place to start on your search for target behaviors is your accident/injury data. What are the most common injuries for your group? If your group has not had many injuries, what about other groups like yours within or out-side of your company? There are lots of great statistics avail-able to tell you what the most common injuries are in cer-tain types of jobs. In many companies strains and sprains are very common as well as injuries from slips and trips. These are great general categories to start with. The next step is to identify the underlying behavioral root cause for injury. You may wonder if doing so will leave you with a very long list of behaviors to monitor. In fact, you'll find that many injuries are caused by just a few behaviors. The Pareto Principle holds true in safety. This principle states that 20 percent of the potential sources of a problem cause 80 percent of the problem. In other words, 20 percent of the at-risk behaviors in a work environment cause 80 percent of the injuries. We have found this to hold true for hundreds of companies. This is good news. If we can identify the 20 percent of at-risk behaviors and turn them into safe habits, we will eliminate the majority of injuries!

Keep in mind that a behavioral root cause may be a barrier such as insufficient training, an unsafe condition, or a poorly written procedure. We consider these to be *behavioral* root causes because, ultimately, someone's behavior (or several people's behavior) created or maintained the barriers. Thus, such barriers can be addressed with the BBS process. This is usually a place where management and/or the safety professionals step in to pinpoint the behaviors needed to address the barrier and ensure that the correct individuals follow through and make the change. Sometimes the behavioral root cause will occur at the frontline level. Procedures are in place but not followed, personal protective equipment is not worn, or machine guards are bypassed. In such cases, the pinpointed behavior is worked on by the frontline groups. Sometimes the solution lies in behaviors at both the frontline level and at the management level. For example, an old piece of equipment may pose a safety hazard. Management behaviors include budgeting for and purchasing a new, safer piece of equipment. However, there is likely to be a time lag before the new equipment arrives. In that case, the frontline group that works with the equipment should focus on developing the safest behaviors to work with the old equipment so they stay safe while waiting for the new equipment.

If your group has had very few injuries over the last several years, you won't have accident/injury data, so you may need to look at near-miss data for the same information. Don't wait until you have an injury to start the process. BBS is about *preventing* injuries. If you haven't had injuries, then look at the most common near-misses and identify the behavioral root causes of those near-misses. If you don't have a good near-miss reporting system, then you may need to rely on observation to identify at-risk behaviors. Most frontline employees can quickly list at least 10 at-risk behaviors that they see themselves and their peers engaging in each day. Frontline employees are a very reliable source for selecting behaviors that are truly important to safety. For example, you may notice that many workers don't wear their hard hats when they get into their trucks despite the fact that the company rule is to wear a hard hat at all times. This is not a good candidate behavior for BBS since it is unlikely that anyone will experience a head injury when in a vehicle (it probably isn't a great company rule either). Select behaviors that really are going to keep people safe even if it takes longer to figure out what those behaviors are.

Specific and Observable Pinpoints

Once you have identified some behaviors that are important to safety, it is critical that those behaviors be *pinpointed*. Pinpointing is the process of describing behaviors very specifically. Why is pinpointing important? First, pinpointing ensures that everyone knows what to do and what not to do. Second, pinpointing makes counting the behavior easier, more accurate, and more consistent. Third, pinpointing increases cooperation and coordination because of common understanding. It is very important that there is no room for disagreement over behaviors. The purpose of BBS is not to set people up to judge one another, but rather to come to a common understanding of the correct way to do something and then work together to improve. If behaviors are not pinpointed, there is room for disagreement and thus argument. Careful pinpointing removes the judgment from BBS and allows people to work together positively.

The first step in pinpointing is to focus on observable behaviors. Behaviors are observable things that people say or do. If *you can't take a picture of it or record it with a tape recorder, it probably isn't a behavior.* The second step is to select behaviors or small tasks (a collection of behaviors) that can be observed in a few minutes or less. Here are some safe behaviors that fall into this category.

- Lift with knees bent and back straight.
- Honk forklift horn at intersections.
- Put on gloves when mixing chemicals.
- Put on face shield before grinding.

You want to avoid large tasks that take many minutes or hours (such as following a lengthy procedure for cleaning up chemical spills or cleaning an entire work area) because the multiple, complex behaviors involved in such tasks are too difficult to observe. If you do want to work on a larger task, simply break it down into its component behaviors and work on those separately.

Reliability

After you have identified observable behaviors, make sure those behaviors are stated in such a way that if two or more people observed an individual doing the behavior(s), the observers would all agree with the same count. For example, if three observers watched two employees mix chemicals, they would easily agree on whether those two employees did or did not put on their gloves. When there is agreement, we say the behavior is *reliable*. An example of a poorly stated behavior is, "Use good posture when typing." Following that directive, if three people observed someone typing, they would probably have three different counts. What is good posture? Many people define that differently. Make sure the behavior is defined so clearly that there is no room for personal interpretation. Make the behavior's definition as black and white as possible and the BBS process will be easier and more effective.

Management and Supervisory Behaviors

So far we have focused on safe behaviors for frontline employees. What about supervisors, team leaders, managers, and executives? They all engage in many behaviors daily that influence the safety of those who work for them. For example, they make decisions on whether to spend money on new, safer equipment, on when and how much safety training will take place, and they make clear where safety is on the company's priority list. They also engage in behaviors that influence their own safety. Working on a computer seems like a safe task that is unlikely to hurt anyone. However, one of the fastest growing categories of injuries is carpal tunnel syndrome and other repetitive strain disorders associated with using computers. Many managers also spend a great deal of time traveling to various company locations, so safe driving behaviors are important. Employees who do not work on the frontline still need to work on developing safe habits to keep them safe whatever the kind of work they do.

Supervisors, managers, and executives should develop and use a scorecard just like the employees on the frontline. The scorecard should include safe behaviors related to the work they do as well as support behaviors. For example, managers might join a group of office workers to work on ergonomic behaviors and other office-related safety behaviors. Support behaviors are those that help support the BBS process and safety in general. Here are some categories of support behaviors.

- Ensure timely removal of barriers to safe behaviors.
- Provide the time and resources for employees to implement BBS.
- Assess the impact of company systems on safety.
- Reinforce behaviors related to BBS at all levels.
- Discuss safety first at every meeting.

- Track the safety work order process and ensure timely correction of problems.

Chapter 13 includes further information about the behaviors of supervisors and managers. The important point, for now, is that supporting safety is simply a series of behaviors, thus it can be improved just like any other behavior using the BBS process. This kind of support is a critical piece of any BBS process. Management support must be clearly specified (pinpointed), measured, and reinforced. Without a positive accountability system to ensure this happens, the probability of success dramatically decreases.

Summary

- Selecting the right behaviors is critical to achieving the ultimate goal of reducing injuries.
- When possible, focus on behaviors that have resulted in injuries or near-misses.
- Clearly state the behaviors so they are easy to observe and so that all observers will agree on the counts.
- Support behaviors should also be pinpointed so that supervisors, managers, and executives know exactly what to do and can be measured and reinforced.

Did You Know?

- The four most dangerous steps on most staircases are the two at the top and the two at the bottom.

- One tsp. of liquid nicotine or 1/2 oz. of pure caffeine are considered lethal doses for a 150-lb. man.

- In Australia, in a recent 10-year period, 380 people were killed or injured by lightning while talking on the phone during thunderstorms.

- In one day, an average typist's hands travel 12.6 miles.

- Wanting always to be prepared, a Cincinnati man wrote to Washington for a copy of the U.S. Government publication #15.700, *Handbook for Emergencies*. Shortly thereafter he had his first emergency: he received 15,700 copies of the handbook.

From Richard Hawk and Company, "Safety Stuff" Newsletter.
Reprinted by permission of the author. www.makesafetyfun.com

CASE STUDY

Behavioral Safety Yields Expected (and Unexpected) Dividends

Dirt roads, Wild West weather conditions, and miles of rolling prairie compose the business setting for a large oil company's Wamsutter and Moxa Arch natural gas fields near Wamsutter, Wyoming, and Granger, Wyoming. The sites contain a total of 700 producing wells, maintained and operated by people who travel the rough terrain of the fields and the 110 miles separating the two locations. The Moxa Arch field (30 miles long and 15 miles wide) and Wamsutter field (40 miles long and 25 miles wide) make up the largest natural gas producing operation in Wyoming and one of the largest in North America. The company continues to drill new wells on the locations at a fairly rapid pace, providing the gas that heats many Midwestern homes. "People work by themselves with lots of travel," said Don Burkhart, safety engineer. "Right now we drive close to a million miles a year among all the workers. I don't think we'll ever find a truck that holds up under those conditions."

Work that's tough on trucks must also be wearing on humans, but this operation holds an impressive 16-year record of no lost-time accidents. Burkhart explained that, in the past, he and Doug Miller, operations center foreman, discussed the reasons for the sites' commendable safety record. Some of that record could easily be attributed to the organization's diligence in maintaining rigorous safety standards and to the safety initiatives the organization participated in over the years. Even so, on a natural gas field, any work that creates heat, sparks, or fire can spell immediate and deadly disaster in the form of raging flames or severe explosions. As sticklers for detail, the two worried, "Could even a small percentage of that record be due to good luck? Are we missing anything that could come back to bite us?"

Burkhart and Miller agree that nonbehavioral safety initiatives might take an operation to a good safety level, but not to a level that minimizes, even eliminates, near-miss risks. The two reviewed and researched many safety approaches and concurred that behavioral methods were the next step toward excellence. Therefore, when their corporate headquarters selected a behavior-based safety (BBS) approach, they were willing to give it a try.

"During training, people looked at the manual and said, 'This won't work; it's way too simple,'" said Burkhart. "We found that part of the beauty of the program is its simplicity." The BBS process involves pinpointing several specific safe behaviors, observing and collecting data, and providing positive recognition and reinforcement as the behaviors progress toward habit. Only when performance of a behavior reaches habit can it be replaced with another pinpoint. Meanwhile, the habit behavior is maintained through routine, but not as frequent, monitoring and reinforcement.

The site management realizes that the independent working nature of the business requires somewhat of a hands-off approach and they view BBS in the same light. "In effect, we turned the people loose with the process," said Burkhart. "They've really taken to it which is one of the successes that surprised us a bit."

Contractor Advantage

Because of another characteristic of the natural gas business, a pivotal decision was made to train contractors in the BBS process. Burkhart explained that the majority—roughly 60 percent—of the fieldwork is performed by contractors. Contractors include the people who deliver pipe, drill and complete wells, operate heavy equipment, or service and build well locations. On this site, hours worked per year by full-time company employees total approximately 120,000. The number of contractor hours per year? About 240,000!

"We look at all people as employees if they work at our facility. It concerns us if we have a contractor accident. It shows us there is something in the system that isn't working," Burkhart said. The inclusion of contractor personnel in the BBS process has resulted in some anticipated positive changes, but also in some unexpected advantages for both the company and the contracting businesses.

"It is terribly difficult to find people who will work in heavy labor," Burkhart explained. "For a while, one of our subcontractors had a 300 percent turnover. Now some of the contractors are realizing, especially with work behaviors, if they run a good safety operation, they get a discount on their worker's compensation program and have more to offer employees." Many contractors retain a core group of long-term employees. The contractors have discovered that those employees have taken to behavioral safety and enjoy the recognition for their participation—even if the recognition is simply a positive word or a cup of coffee. "Many contractors have concentrated on getting 100 percent behavior with their central group of people and letting that central group push the behaviors to the newer employees," said Burkhart. Traditionally, contractors have struggled to get new employees to consistently meet the most basic safety requirements, such as wearing hard hats and safety glasses. Yet, those behaviors quickly rose from 50 to 75 percent once the behavioral process began.

One contractor encourages employees to periodically videotape operations, and then give one another feedback during the film screenings. The banter during the viewing is constructive and fun, something that the workers see as an enjoyable, even entertaining, form of useful feedback. "The contractors have really given me a good feeling the way they've used BBS," said Burkhart. "I think BBS gave them the push they needed to start some of these things."

Behavioral Thinking

The people at this Wyoming operation soon discovered that examining performance from a behavioral perspective almost inevitably expands to every aspect of the workday. "Even in areas that we aren't scorecarding, people have taken the positive reinforcement process and started adding to it, doing things on their own. I think the phrase is 'thinking behaviorally.' It pays dividends," said Burkhart.

For example, decision makers at the operation began to look at how work conditions might impact behavior. Several contractor accidents occurred during overtime work, often taking place in adverse weather conditions. (In the winter months, air temperatures at the two fields often dive to ranges of 0 to –20 degrees Fahrenheit with wind chill temperatures from 0 to −50 degrees Fahrenheit.) Following BBS training, the supervision encouraged analyzing, in terms of risk assessment, any work that exceeded 12 hours or was of a non-routine nature such as work conducted after sunset. With the go-ahead to examine operations with an emphasis on behavior, Howard Sanders, completions technical specialist, came up with a safer and more satisfying solution. He expanded the hours for well completion to four full days rather than the usual three days plus a few hours, which involved overtime. "They cut overtime which cut our costs," Burkhart explained. "The employees of the contractor liked it because they got home at a reasonable hour during the week as well as had an additional full day of employment. Also, although it was not one of our pinpoints, we felt that change made the job safer."

Risk Management

In work settings across America, standard procedures are often a topic of contention. Those in the field may think that the procedures create unnecessary work, and those who penned the procedures think that the workers don't need further explanation. Unfortunately, despite the conflicting viewpoints, communication regarding the procedures often never takes place. Instead, workers may ignore steps they see as superfluous, often at risk to their own safety and that of others. Such a situation became apparent at the Wyoming fields when one employee's deviation from procedure caused a potentially hazardous gas flash fire.

After behavioral safety training, one operator's

group picked the difficult procedure involved in that incident—the lighting of fired vessels—as a focus of their BBS implementation.

Fired vessels are heated tanks using large burners and fueled by natural gas. If the vessels, which resemble large boilers, are lighted incorrectly, serious burns, damage to equipment, fire, and natural gas explosions may result. On average, every employee lights a fired vessel at least once a week, a task that is spelled out via step-by-step instructions. Yet, when the group collected data, they found that the procedure was correctly completed only 85 percent of the time.

Without pressure from supervision, the operators' team convened to determine why the potentially dangerous scenario of ignoring procedures happened during 15 percent of the lightings. The group decided the procedure was incorrect; two of the required steps added time without enhancing safety, and therefore, people were prone to omit them. They came to Burkhart with the news. Importantly, rather than reacting defensively, he agreed to examine their claims, walk through the procedure with the group, and consider their suggestions. During the walk-through, Burkhart explained the reason one of the steps was crucial to safe lighting procedure. "They said, 'Nobody has ever explained that, so we'll put it back in,'" said Burkhart. Also, he realized that one of the steps was indeed redundant and did not add to safety. The compromise resulted in a change of the procedure and a newfound understanding on both sides. Everyone learned a valuable lesson about the importance of communication between those who write procedures and those who do the job.

"To me, a very powerful thing is to get people to think, make decisions, and adjust their behavior to changes and conditions in a safe manner," Burkhart concluded. "It's allowing people also to say, 'Wait, we think the procedure is wrong,' and evaluate it for themselves. To me this is the way BBS should work and we see it as a major success. The process helps people think about what they are doing and gives them the tools they need to do something about it."

Savings Now and in the Future

Burkhart recently shared the podium with Wyoming's OSHA compliance director, Wayne Walm, during a presentation to the local chapter of the American Society of Safety Engineers. One audience member asked about the cost effectiveness of the behavioral safety process. Burkhart replied, "According to the national statistics, the cost of the BBS process was roughly the average cost of one disabling accident. If you've prevented one accident, you've paid for the process, but we've prevented more than one accident. And, we will prevent future ones as well."

CHAPTER 9

OBSERVING AND MEASURING SAFETY IN ACTION

For many people, observations are synonymous with BBS. In fact, many people falsely believe that observation is the heart of BBS and is the primary factor in behavior change. Yet, behavioral observations, like all other types of measurement, do not cause behavior change: they simply provide opportunities for feedback and reinforcement; then, when properly delivered, feedback and reinforcement cause behavior change.

While measurement is not the cause of behavior change, it is a critical prerequisite. Precise measures provide us with accurate information on the frequency of safe behaviors. To be effective with feedback and reinforcement we must know exactly how safely people are behaving. As you will recall from our discussion of consequences, the more specific consequences are, the better. It isn't very reinforcing to say, "It seems to me we are a little safer than we were last week . . . good job." By having good measures we can state specifically how much safer we are and have much more confidence that performance is worthy of positive recognition and reward. Another great reason to use measurement is the opportunity to notice small improvements. Most behavior change happens gradually. Often it is difficult to notice a difference from one day to the next or from one week to the next. A sensitive measurement system will allow you to detect those small changes and reinforce them. The more you reinforce the small improvements, the faster the behavior will improve. So as you can see, a good measurement system really is important.

Sampling for Safety

Because BBS focuses on observable behaviors, the simplest way to measure those behaviors is through observation or simple counting. Ideally we would count every instance of each safe behavior we wanted to improve but that is hardly practical. Instead we rely on sampling. Everyone is familiar with sampling during elections. Rather than ask every voter in the country how they are going to vote to predict a winner, pollsters ask a sample of people. If sampling is done well, it can give you a very accurate picture of what is really happening. To get a representative sample of safe behavior, observations should be done every day, several times a day, across all people in the work group. Since we recommend that you focus on just a few behaviors at a time, this is actually very easy to do. People who use these BBS methods state that it takes just a few minutes a day to complete observations; and since this time is spread throughout the day, it really doesn't seem to take any time at all. In fact, observations are usually folded into the normal activities of work. Here is an example of an observation scorecard.

SAMPLE SCORECARD FOR CORE TEAMS

BEHAVIORS	SAFE	AT-RISK	BARRIERS	TOTALS
Wipe up spills immediately	/// /	///	Cleaning materials unavailable; could not leave work station	4/7
Put on gloves and sleeves prior to starting task	/// //	///		5/8
Lift with knees bent, chin up and chest out	/// /	/// //		4/9

Doing an observation is quick and easy. For example, as you walk to get a tool or some paperwork, you might look around and do some observations. You would simply look for the behaviors your group agreed on and put a tick mark under either the *safe* or *at-risk* column, depending on which you saw. (Feedback will be covered in another chapter.) An hour later, sitting on a forklift waiting for another vehicle to clear out of your way, you might look around and do another observation. It takes just seconds.

Another advantage to doing observations periodically throughout the day is that you are more likely to observe people when they are acting naturally. This is important since we want people to do the safe behaviors whether someone is watching or not. Keep in mind: everyone being observed should know that they are going to be observed throughout the day and they must know exactly which behaviors are being observed. BBS is not a game of gotcha! People know which behaviors to engage in and they know they will be observed; they just don't know the exact moment they are being observed. Again, this approach is important because it provides a more precise and realistic measure of the frequency of the safe behavior. If you warn someone that you are going to observe them for the next few minutes, they are highly likely to do everything safely and your numbers will be artificially inflated. Instead, be sure everyone knows they will be observed but let them go about work as usual. The numbers you get from this type of observation will much more accurately reflect the safety of your work area. These numbers are used for positive reinforcement only, so this kind of system never results in people feeling as though they are being spied upon or caught.

Since the ideal way to perform observations is to do quick observations throughout the day, who is in the best position to do those observations? You guessed it . . . peers. The exception to this rule is those people who work by themselves some or all of the time. If you work by yourself, the only one who can do frequent observations of your behavior is you. A later chapter will cover self-observation —a tested and very effective way to use BBS with the lone worker. The remainder of this chapter will focus on peer-to-peer observations.

The Snitch Factor

Initially some people are concerned about doing observations on their peers. They worry that BBS is a peer-policing system or a snitching process. Both words *policing* and *snitching* imply the use of negative consequences. By now you know that is exactly the opposite of the BBS process which focuses on the use of positive reinforcement. Furthermore, BBS is an anonymous process. There are no names on the scorecards so it is impossible to know who did the safe behaviors and who did not. In addition, the data from observations is collected by and summarized into percentages by peers. Thus, no one in management will even see the scorecards. All that management will see is the percentage of time that the group performs a particular safe behavior. Very quickly the reservations about doing peer observations disappear as people realize the company's commitment to anonymity and positive reinforcement.

Barrier Identification

A big part of the observation process is the identification of barriers to safe behavior. Often the observer can see a barrier when doing observations and sometimes barriers are uncovered in the feedback conversation that occurs after an observation (more about this in the next chapter). One utility, for example, had a maintenance group that serviced large vehicles. An important safe behavior for the group was to use three-point contact (making sure that three of four limbs are in contact with the vehicle at all times) when getting on and off the vehicles. Early in the observation process, they discovered a barrier. One of the trucks had too much distance between the cab of the truck and the first step. This resulted in people having to jump from the cab to the first step with a high risk of falling. Through the BBS process this barrier was identified and corrected by adding a second step to bridge the gap. In addition to physical barriers there may be barriers related to lack of procedure or poor procedure or training issues. It is important that the BBS observation process enable you to capture these barriers (note the Barriers column on the sample scorecard) and that a process be developed to deal with barriers as quickly as possible.

Measuring Management Behaviors

So far we have focused on the measurement of safe behaviors by frontline workers. In the previous chapter, we discussed the importance of pinpointing support behaviors for supervisors, managers, and executives. As with the safe behaviors, we must also measure these behaviors if we want to improve them. The best approach to measuring support behaviors is to have supervisors, managers, and executives track their own performance on these behaviors. Then a positive accountability system can be established in which each self-scored support scorecard is turned into that individual's boss on a weekly basis. If progress has been made, this will lead to positive reinforcement. In addition, it is important for an independent party to periodically check with the frontline workers to make sure they are getting the support they need. If a supervisor is scoring high on his/her support scorecard but his/her direct reports say they are not getting the support they need, then something is amiss. Often this is a matter of finding the correct support behaviors and that may not happen perfectly the first time out. Chapter 13 provides more information on the role of management.

Summary

- Measurement is important because it helps us see improvements, even small ones.

- Measurement doesn't change behavior. Measurement simply enables the use of feedback and reinforcement—the most important elements for changing behavior.

- Since it is impossible to track every instance of a targeted safe behavior, we use sampling.

- A good sample must be large enough to be representative; therefore, it is important to observe as frequently as possible.

- The observation process is anonymous and *never* results in punishment.

- Make sure that all employees know what the target behaviors are and that they will be observed at some time. If observations lead to positive reinforcement, no one will mind being observed.

- Support behaviors are measured through a self-scoring system and by periodic checks with the frontline workers who are being supported.

Did You Know?

- Statistics suggest you have a much better chance of living without back problems if you live in the north-eastern United States. Your risk increases by almost 45 percent if you live in the West.

- In terms of fatalities per trip, buses are about 100 times safer than automobiles.

- In 1998 the National Park Service reported 146 deaths and 6,638 search-and-rescue efforts.

From Richard Hawk and Company, "Safety Stuff" Newsletter.
Reprinted by permission of the author. www.makesafetyfun.com

CHAPTER 10

HOW AM I DOING?
THE VALUE OF FUNCTIONAL FEEDBACK

Once you have collected some data using the observation process, it is important to provide feedback to the performers about how they are doing. *Feedback is defined as information about your performance that helps you improve your performance.* So, feedback isn't just numbers; it has to be useful in helping people improve. Business is filled with numbers and most employees are accustomed to hearing about or seeing many different numbers each day (production numbers, quality scores, customer survey results), but how much of this is truly feedback in that it helps employees perform better? Much of it is simply information.

Feedback is extremely important for behavior change. In fact, we can't change without feedback. Whenever we attempt to change a personal behavior, we look for ways to get feedback on our performance. If you are attempting to improve your golf swing, you may hire a golf pro to give you feedback. If you are attempting to lose weight, you step on a scale each week to get feedback on your progress. It is important to note that feedback alone does not change behavior. Feedback is necessary but not sufficient for behavior change. Consequences are required for behavior change. Feedback without consequences will lead to short-term behavior change at best. If feedback is not strongly associated with any good or bad consequences, it will lose its effectiveness.

In BBS we rely on several different types of feedback: group graphic feedback and individual verbal feedback (positive and constructive).

Group Graphic Feedback

Graphic feedback is effective because it allows performers to see their progress over time. Since we do not collect data on an individual basis, there is no way to create *individual* graphic feedback. The only exception to this is the lone worker. The lone worker who does self-observation exclusively can graph his/her progress individually if he/she chooses. In most cases, people do not create an individual graph since the scorecard itself provides the feedback on how the lone worker is doing. As you see more tick marks in the safe column, you know you are getting better. For both the lone worker and those doing peer observations, the most common approach is to create graphs of *group* performance as well as provide verbal, group feedback at daily or weekly meetings. Each behavior should have a graph showing percent safe over time. This gives the group something to rally around.

Lift With Knees Bent and Back Straight

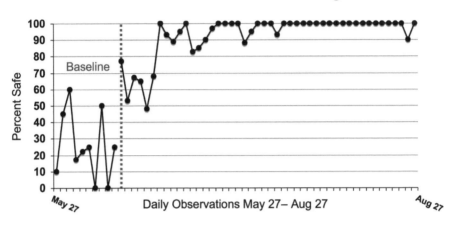

Daily Observations May 27– Aug 27

Tips for Graphic Feedback

LABEL THE GRAPH CLEARLY

A good graph is quickly updated and easy to understand. Put the pinpointed behavior on the top of the graph. The y-axis should be labeled *percent safe* and the x-axis should be labeled *days* (*weeks* in some cases).

ESTABLISH A BASELINE

Feedback is only effective if it tells us how we are doing relative to our past performance. A baseline is a measure of performance, in this case percent safe, before any attempts to change. By collecting baseline data, performers can see their improvement more readily and can continually look back to where they were before beginning the safety process. This is particularly helpful when the group is struggling with those last few percentage points. It is easy to get discouraged, but a quick glance at the baseline will provide lots of encouragement as groups see how far they have come.

ESTABLISH A GOAL AND SUBGOALS

Just like we need to know where we were (baseline), we need to know where we are going. In most cases in BBS your goal should be 100 percent safe. With most safe behaviors anything less than 100 percent means someone could get hurt. Getting to 100 percent is a big challenge, especially if you start at 30-40 percent. Now you understand that future and uncertain consequences are less effective than more immediate and more certain consequences. So, you can see that waiting until the group reaches 100 percent before you recognize improvement is not a good idea. It may be many months before that happens and reinforcement needs to be much more frequent than that. Instead, set subgoals along the way. Subgoals are simply times to stop and celebrate accomplishments. The more you do this, the faster the progress. Make subgoals part of your feedback so that everyone has a short-term objective to strive for and for which they can feel good about accomplishing.

Individual Verbal Feedback

Individual verbal feedback is an important tool in developing safe habits. Since observations are anonymous, the only person who can give individual feedback is the observer. After the observation is complete, there is simply a tick mark on a scorecard and it is impossible for anyone to go back and provide individual feedback using that information alone. Thus, the observers should provide feedback at the time of observations. Now, while it might seem like the observers should give feedback every time they do an observation, keep in mind that observers are doing 20 or more brief observations a day. Not only would it be impractical to ask them to give feedback after each observation, it is actually undesirable. Hearing many times a day that you have done a particular behavior safely would become tedious and probably annoying. We recommend that observers deliver feedback when they believe it will have the most impact.

Listed below are some suggestions for the best times to deliver feedback.

- When a performer is in imminent danger (*always* intervene)
- When a performer does the safe behavior for the first time
- When a performer has made recent improvements
- When a performer has been struggling to change an unsafe habit
- When the safe behavior is done at a particularly difficult time (or situation)
- If you notice barriers to the safe behavior
- If you believe the performer is unaware of doing the at-risk behavior at the moment of the observation

Remember, always be as specific as possible. For example, you might say, "Every time I've observed you today you have been wearing your gloves; you're really making progress."

Positive and Constructive Feedback

Of course, we recommend a very heavy emphasis on positive rather than constructive feedback. We have already made the case for focusing on the use of positive reinforcement versus punishment or negative reinforcement. Whenever possible it is a good idea to focus on the use of positive feedback since that becomes a powerful antecedent for using positive reinforcement. If you focus on constructive feedback, the antecedent is for negative consequences, not positive ones. Of course, there are occasions when constructive feedback is not only desirable but necessary. If someone is in imminent danger because of his/her behavior or if someone isn't aware that he/she is doing the at-risk behavior, then constructive feedback is appropriate. Just keep in mind that if you maintain a high rate of positive feedback and positive reinforcement for safe behaviors, people will be more open when constructive feedback is necessary.

Tips for Verbal Feedback

USE "I" STATEMENTS. Being on the receiving end of feedback often makes people a little defensive. By starting your feedback with *I* instead of *You,* the receiver is less likely to get defensive. "I noticed that…" rather than "You did . . . ." *I* statements also help you demonstrate you are not just following a process, but that you really care about an individual's safety. For example, "I am concerned when I see . . . ."

DISCUSS ONLY THE BEHAVIOR(S) YOU OBSERVED. Avoid generalizing and saying things like "You are really unsafe." Just describe what safe or at-risk behavior you saw. The more specific the feedback is, the less it feels like a personal attack.

USE APPROPRIATE TONE AND BODY LANGUAGE. Your tone of voice and body language should demonstrate concern. Avoid sarcasm and sounding accusatory.

ASK IF THERE ARE BARRIERS TO THE SAFE BEHAVIOR. If an at-risk behavior is observed, it is important to ask the performer if there were barriers that made it difficult or impossible to do the safe behavior. Make note of any barriers identified so they can be removed, if possible.

INCLUDE AN IMPACT STATEMENT. Unless it is obvious, the observer should point out the impact of the safe or at-risk behavior. For example, "Twisting at the waist instead of pivoting with your feet can really hurt your back over time."

ACKNOWLEDGE PROGRESS. Include a positive statement if the performer has been getting better. Again, the more specific the better. For example, "The last four times I observed, you stood out of the line of fire when opening that valve. You are really improving on this behavior!"

DO NOT ARGUE. Some people react negatively to feedback (especially constructive feedback). It is critical to avoid arguing. Acknowledge their feelings, state your purpose (to keep them safe), restate what you observed, and walk away. For example, "I can see you are upset. I just want to be sure you don't get hurt. So I wanted to tell you that I saw you operating the saw without your glasses." Then walk away.

As you can see there are many choices when it comes to feedback. A good BBS process includes all of them. Remember: feedback and reinforcement are the tools for changing behavior so the more often people receive both and the more varied ways people receive both, the better. As long as the focus is more on the positive, you'll find that feedback is welcome.

Summary

- The more you learn about feedback the better; but if you follow the tips provided in this chapter your BBS feedback will help you achieve the goal of improved safety.

- Verbal and graphic feedback are helpful, so both should be included in BBS.

- While constructive feedback is necessary at times, try to use more positive feedback.

- To make feedback most effective be specific, establish a baseline, and establish subgoals to mark achievement.

- Remember, we all want to know how we are doing as often as possible and as specifically as possible.

Did You Know?

- About half the people who drown are alone at the time.

- The average American is involved in six motor vehicle accidents during his or her lifetime.

- In the Old West many more cowboys died from falling off their horses than were killed by bullets.

From Richard Hawk and Company, "Safety Stuff" Newsletter. Reprinted by permission of the author. www.makesafetyfun.com

CHAPTER 11

POSITIVE REINFORCEMENT: THE "WAY-TO-GO!" CONSEQUENCE

Why do we pay to go golfing, but have to be paid to come to work? The answer is directly related to the amount of positive reinforcement involved with each activity. Some activities, playing a sport for example, offer so much reinforcement that we'll spend as much time as possible doing those activities and trying to become better at them. So why can't we inject more enthusiasm into work by building in more positive reinforcement? We can.

As explained in Chapter 4, the receiver of the consequence *(not* the person who offers the consequence) determines whether that consequence is positive or negative. For example, just because you want your son to have a new deck of flash cards as a reward for doing well on a math test doesn't mean that this consequence will function as a positive consequence. (In fact, it probably won't! Who wants flash cards as a reward?) With BBS, as in any behavior change initiative, everyone involved must also remember that *positive reinforcement is any desirable consequence that follows a behavior and increases its frequency in the future.* To fully understand why positive reinforcement plays such an important part in behavior-based safety, one should be aware of the types of positive reinforcement available to us.

Self-Reinforcement

If you really think that a behavior is important and you begin to see evidence that you're getting better at that behavior, you may say to yourself, "I'm improving. Good for me!" But sometimes, even often, we are our own harshest critics. Achieving safe habits on the job would be a cinch if one could simply march out and proclaim, "Okay everyone. When you put on your safety gloves today, tell yourself you're doing great. From now on, I want everyone to self-reinforce for that behavior." Well that would do the trick, but that's not going to happen. Sure there are those people who have learned to be self-reinforcers, consistently giving themselves positive self-talk and recognizing their own merits, but this isn't a characteristic that can be counted on for an entire workforce. Some people are good at self-reinforcement and some are not, so we can't rely on self-reinforcement as a viable and consistent way to change behavior. Also, people sometimes reinforce themselves for unsafe behaviors instead of safe behaviors (such as feeling proud of lightening-fast reactions that allow them to avoid accidents when speeding). In other words, they tend to go for the *natural reinforcers* associated with such behavior.

Natural Reinforcers

Natural reinforcers are essentially reinforcers that no one engineers, but that happen as you engage in a behavior. For example, in golf, when you hit the ball and you hit it well, several natural reinforcers occur. If you hit the ball correctly, the ball goes far and straight: a natural reinforcer. When you hit that ball just right, you don't even have to look at it. You can tell you've made a good drive because of the pitch-perfect sound emitted at the moment of contact between ball and golf club. The same is true with baseball. Seasoned players don't even have to look at the ball to know when they've hit a homer. Once again, the sound alone is a natural reinforcer. Add to these natural reinforcers the fact that people may be congratulating you on your shot or cheering your hit and you've added *external reinforcement* to the game. The amount of positive reinforcement received determines the difference between work and play.

External Reinforcers

One person in a large manufacturing plant pointed out that she consistently wears her safety gloves because she hates to get her hands dirty. She doesn't have a problem with that behavior because for her there is a natural positive consequence: she keeps her hands clean. Others in the same plant have cut their hands when they didn't wear their safety gloves, so they now consistently wear them. However, there are many people who don't care if their hands get dirty and who don't think they're going to cut their hands. Those are the situations that require some positive external reinforcement. After all, if putting on safety gear felt just great, then people would do it consistently. Maybe they wouldn't even want to take the safety gear off!

Unfortunately, most of the naturally occurring consequences surrounding safe behavior are actually punishers. It's uncomfortable to wear a lot of personal protective equipment; it's a hassle to put on safety gear; it's time-consuming to take all of the by-the-book precautions. Those are all natural and negative consequences that discourage us from doing the right thing. They are barriers to safe behavior. So, how do we counter these natural punishers? When possible, we eliminate them. For example, more comfortable PPE is purchased and safety procedures are streamlined to make them easier to do. In BBS we first work to alter the natural consequences to encourage safe behaviors. When that isn't possible, the only consistent and effective option is to provide external reinforcement for safe behavior in the form of *social* or *tangible* reinforcers.

Social Reinforcers

A social reinforcer is any positively reinforcing interaction between people. Tangible reinforcers are items of physical or financial value. Which type of reinforcer should we attempt to provide most frequently? It may surprise you to learn that social reinforcers are far more effective for bringing about behavior change.

Social reinforcement is a way of saying, "We don't take you for granted. We notice and appreciate what you're doing." If we want to make sure that people change their behavior, personal interaction is very powerful, whereas using *only* tangible reinforcers has proven to be widely ineffective. Also, social reinforcement is something any of us can do at any time, which is another of its values. When it comes to external reinforcers, the ideal mix is to use mostly social reinforcers with an occasional but symbolic tangible item.

Many of us think of social reinforcement as verbal praise such as "thank-you" or "good job" and of course verbal praise is a common form of social reinforcement. However, you must be careful when using face-to-face or public recognition. Experience has shown us that most people don't like to be recognized in front of large groups of people, but prefer to receive recognition privately or within their own small work groups. There are exceptions to every rule, which is why it is important to consider each individual before attempting public reinforcement.

Keep in mind at all times that the giver's intention does not matter in reinforcement. You may recognize a desired behavior with words or an item that you assume is a positive reinforcer. However, if the behavior does not happen more often following your reinforcement attempt, then it was not a reinforcer. In fact, if the behavior happens less often or stops, your intended reinforcer was actually a punisher. For example, one company that began a BBS initiative, readily embraced the idea of positive recognition and reward, but they sometimes blundered at reinforcement attempts because they didn't first get to know the desires of the workforce. Every morning, supervisors and managers conducted a meeting from a stage at the front of the company warehouse. During the meeting, they talked about daily production and so on, but they also publicly recognized individuals and teams for safe performance by calling them up on stage. One day during the event, one long-term employee was overheard saying, "If they ever call me up on that stage I'm walking out of here and I'm never coming back!"

This company's management believed it was doing a good thing; but through their recognition efforts, they risked alienating, even losing, valued employees. Furthermore, those who cringed at the thought of being on stage may have gone out of the way to avoid being caught doing a safe behavior! Ask people to fill out a questionnaire or answer a survey on the types of social reinforcers they enjoy most. Otherwise, you could be sabotaging your best intended efforts and compromising the success of your organization's BBS process.

Examples of Social Reinforcers

The power of social reinforcement is that it is available at all times and anyone can offer it. Peers are usually in the best position to reinforce each other. Peers often know better than supervisors who is doing what, who is getting better (and who is not), and they also know who likes what kind of social reinforcement. This doesn't mean that managers and supervisors shouldn't reinforce employees, but employees can reinforce peers more frequently and with BBS are asked to do so. Also, none of us should forget that our managers and supervisors enjoy positive recognition.

Following are examples of ways to socially reinforce one another.

PRAISE. Any statement acknowledging that a performer did something right is usually reinforcing.

GESTURES. Small gestures that indicate you saw the performer do a safe behavior are very useful in BBS. When someone is trying to develop a safe habit, any indication of improvement is reinforcing. Thumbs-up, smiles, nods, or a pat on the back can all be effective at increasing safe behavior.

NOTES. Written notes, cards, e-mails, or sticky-notes are effective ways to reinforce performers who are uncomfortable with public fanfare and recognition.

ASKING FOR INPUT. For most of us, being asked for advice or "How would you do it?" is an affirming and positive experience. Of course, if you ask for someone's advice or opinion, you must sincerely want to hear it. Insincerity kills any attempt at reinforcement.

USE HUMOR. An element of fun can be added into any workday. As long as humor is used appropriately and everyone enjoys a laugh, it creates a memorable experience. With some imagination, humor and fun can easily be added to meetings and celebrations to add to their social reinforcement value.

OFFER MORE AUTONOMY AND FREEDOM. Frontline personnel enjoy the freedom to choose how they are going to tailor the BBS process to fit their needs in areas such as observations, data collection, graphing, goal setting, pinpointing, reinforcement, and celebrations. That autonomy and freedom, in and of itself, is a highly valued reinforcer.

ATTENTION. Don't ever underestimate the effectiveness of paying positive attention to somebody's behavior. Most people understand the importance of giving time and attention to their spouse, children, extended family, and friends. Your time and attention can be just as powerful at work. Pay attention to the good things that people do. If we don't pay attention to the good things, we have a tendency to pay attention to the bad things. It may sound simple, but it's a crucial mental shift we have to make.

Discovering Social Rewards

Valued, trusted, competent, intelligent, respected, appreciated . . . this is a list of things that most of us like to feel at work. Take a look at the list and pick the three that are most important to you. Think about how people at work have made you feel that way. Come up with one example for each of your three choices. You will probably find that most of the things that made you feel valued, trusted, competent, intelligent, respected, and/or appreciated were social interactions.

With social reinforcement you are focusing on desired

behaviors that people do every day and trying to increase those desired behaviors. You don't have to gather the troops for a round of applause for every single behavior. That would be overkill and potentially punishing. However, most people who are trying to change or develop a habit are reinforced (and their behavior happens more often) when they know that somebody else notices the change. All you have to do is nod, give a thumbs-up, or make a positive, quick comment. Anything that sends the message, "I see you trying to do the right behavior" really has an impact.

Frequency Is Key

Have you ever tried to lose weight? If so, you may have noticed that people tend to provide you with positive reinforcement once they see the results, once you've finally lost the weight, right? "Hey," they comment. "You lost weight! You look good!" Of course, this recognition of your hard-earned results is great, but do you know when you need that support the most? You need it the most when you're standing in the cafeteria line making decisions: extra cheese or not, a cookie or an apple. That's the time when it would really help to hear somebody say, "I saw you walk past the dessert section; you really are sticking to your diet." Standing in the cafeteria line, resisting dessert at the restaurant, not having that extra piece of bread . . . those are the times when a little bit of reinforcement goes a long way. Those are the moments that enable you to get to the place where people say, "Wow. You're looking good." After all, if you don't choose multiple times a day to eat the right thing, then you're never going to lose weight!

The same is true of behavior-based safety. We can't wait until the organization's accident rate has gone down to provide needed recognition. We must reinforce people for the safe choices they make throughout the day. We don't have to do that with balloons, fanfares, big rounds of applause, or by bringing them up on a stage. We just let them know that we saw them make a safe choice.

Tangible Reinforcers

Tangible reinforcers—items of physical or financial value—can range from something small like a cup of coffee, a cookie, or a pen with the company logo to a cash award or gift certificate. Caps, jackets, and T-shirts can also be reinforcing if paired with a meaningful event. Some companies have used tangible items that relate to the work done by those receiving the tangibles, such as tape measures or pocketknives. One organization even gave out safety gear as a reinforcer. The work team was a group of young people who resisted wearing their safety glasses because they didn't feel they looked cool when wearing them. When the group began to improve their safe behaviors, they each received a pair of slick, funky eye gear that didn't subtract from the group's fashion statement.

The most commonly used tangibles are catered food events or celebrations, because food is definitely a tangible that most people enjoy. Another great thing about using food as a reinforcer is that it is really a blend of social and tangible if it involves getting people together to talk, eat, and celebrate accomplishments.

Do people like tangible reinforcers? Yes. The rule to remember, however, is this: tangibles are a great thing to do if you don't overdo. After all, if you receive a mug, a key chain, a ball cap, even a donut every day, after a very short while those items hold no meaning, especially if little social reinforcement is involved. In an inspirational video, "The Sid Story," a hardnosed supervisor learns the power of positive reinforcement. In the film, Sid astonishes his crew by bringing in his wife's homemade pumpkin bread when the group achieves a goal. The reinforcer proves incredibly effective not because of the pumpkin bread, but because Sid personalized the recognition by asking his wife to bake it and he adds, "The pumpkin bread is from me and my wife." If Sid had purchased the pumpkin bread and plunked it down in the break room, who would have cared? This is but one example of the importance of pairing any tangible with social meaning.

For the BBS process to be successful, an organization should keep as much reinforcement going as possible. It isn't practical to do this using tangibles throughout the day. Not only do the tangibles lose their meaning, few if any companies could afford it. Also, it is not the goal of the BBS process to turn managers and supervisors into work-related Santas, but instead to reinforce safe habits through recognition and reward.

Certificates, plaques, and trophies are frequently used tangibles that may or may not have any discernible effect on performance. Those same items provide a good lesson about positive reinforcement. Many of us own a treasured plaque, certificate, or trophy that we received many years ago and will probably keep forever. Just as many of us probably have such items that we threw away, gave to the kids, or stashed in the attic. What's the difference between two trophies, one that you value and one that you do not? The difference lies in the memory about how and why you received this item. When we look at a valued certificate, trophy, or plaque, we think about what we did to earn it and typically there is some self-reinforcement involved. Such is the perfect tangible reinforcer. A tangible reinforcer should be one that every time you look at it, you think about what you did to earn it and/or the words said by the person who presented it, and you relive the good feeling.

The Dollars Dilemma

We can choose one of two paths in our approach to tangible reinforcers. One path is to assume that the value of a tangible depends on how much money you spend. Those who follow this path believe that the more expensive the item, the more reinforcing it will be. This is not the case. If you go down this path you will disappoint, even anger people, and misspend money on a series of tangibles with escalating price tags. You will not reinforce people and you will not get the behavior changes you seek.

The other path is to assume that what makes a tangible item reinforcing is the meaning behind it or what it represents. This is the path of creating memories by providing items that remind people of their accomplishments. You create valuable memories when people look at something and remember what they did and feel good about it all over again. If you choose this path, you may still make some errors, but you will have more success and you will accomplish significant behavior change.

When you deliver tangible items, always consider it your job to make the item meaningful. Whether you're handing out a candy bar or giving out a T-shirt, your goal should be to make that tangible something that reminds the recipient of what he or she did to earn it or what he or she is proud of, and that leaves the recipient with a story to tell.

Remember, we want to make sure that people know we value what they do already and if we want them to do better, we must be willing to reinforce that effort. Tangibles are simply the icing on the cake.

Summary

- BBS focuses on external reinforcement because it is easily changed.

- Both social and tangible reinforcement can be effective but social reinforcement should be used the most.

- Social reinforcement can be used anytime, anywhere by anyone, so by focusing on it we can generate much more overall reinforcement.

- Tangibles are most effective when they have trophy value and generate a positive story to tell in the future, if only to ourselves.

Did You Know?

Some sure-fire ways to make someone feel bad instead of good:

- Saying something sarcastic after one person has given positive recognition to a co-worker.

- Following positive recognition with the word "but." ("Your report is complete, but it's five minutes late.")

- Making a joke or a disclaimer. (For instance, following a positive comment with, "I guess now you'll really have a swelled head, huh?")

- Asking for more. (Resist the temptation to say, "Keep it up!")

(Source: Recognition Bloopers and Blunders from *You Made My Day: Creating Co-Worker Recognition and Relationships* by Janis Allen and Mike McCarthy.)

CASE STUDY

Out-of-the-Box Leadership

(Also see "Giving Safety a Lift at the Grande Prairie Sawmill" Chapter 5.)

For the kickoff of the Olympic-themed safe lifting intervention at the Grande Prairie sawmill, Doug Chappell, dry side manager, and his associates drummed up the courage to step out of their comfortable management personas. At the beginning of the Monday 7 a.m. shift, Chappell and Linda Thomson, team leader, marched as a two-person parade through the sawmill. Chappell carried a sign stating, "Olympics Team Planer Mill Going for Gold" as his associate led the way carrying an Olympic torch replica. As the two passed by, they were completely ignored by the first few people on the line. "At that point I couldn't help but think, 'Boy, is this the right thing to do?'" said Chappell.

Then a worker farther down the line grinned and gave Chappell a thumbs-up. Soon, others began to respond with encouraging comments. "Overall, the reaction was very positive," Chappell said, then added, "Leadership really needs to get outside of their own comfort zone and that was definitely outside of my comfort zone!"

Breaking the ice gave the leadership team at Grande Prairie the courage to do more out-of-the-box thinking, such as brainstorming a variety of celebrations. When the mill workers reached the bronze medal performance standards set for safe lifting, the managers served them "Bronze Medal Sundaes," covered in bronze butterscotch sauce.

When the two shifts reached the silver medal milestone, Chappell personally baked cupcakes and decorated them with silver icing. "I made a heck of a mess in my wife's kitchen," he confessed. After labeling the desserts with the words "Silver Medal Cup (cake)," he served them to the crew. "What's a cupcake?" he said. "Yet, several people came up and said, 'It's really neat that you did that for us.'"

When the mill team reached the gold medal performance level, leadership gave each person an enamel mug embossed in gold with a logo of an Olympic flame and the words, "Lifting Olympics Team Planer Mill Gold Medal Performance."

"We hope the cup serves as a memory anchor. If somebody sees the cup sitting in a kitchen or at a workstation and asks about it, then the owner of the cup has a story to tell," Chappell commented.

After the planer team achieved gold medal standards, the mill managers wanted to reinforce maintenance of safe lifting behaviors. The managers came up with the idea of drawing four employee names and an activity for each for every five consecutive days that the gold (98 percent or more) safe lifting behaviors continued. The rewards for the winning employees included such things as extended lunch hours, lunch with the manager, leadership replacing the employee on the line and giving him/her two hours of free time, or a manager washing the employee's vehicle. "I've learned that folks like almost anything that might be viewed as leadership being humiliated," Chappell said with a laugh. Consequently, when the managers washed an employee vehicle, they did so right outside the lunchroom windows at noon so the whole crew could watch. "People enjoyed it so much they came out with their cameras and took pictures of us washing the cars and trucks," he said.

With plans to maintain the habit of safe lifting behaviors, the Olympic Lifting initiative ended officially with a closing ceremony and a parade of champions—those who participated 100 percent throughout the program. Two large banners remain in the mill: one stating, "There's Gold In Them There Lifts" and the other proclaiming, "Every Lift Is A Golden Opportunity."

"The banners serve as an antecedent/reminder for the safe behavior and as a source of reinforcement when people here get to talk to others who ask questions about the banners," said Chappell.

The behaviors of the lifting Olympics appear to have observable staying power. Because of the noisy nature of the mill, verbal feedback isn't always possible, but the people at Grande Prairie can still get the message across. "Quite often somebody 50 feet away will see you lift something; they let out a yell and give you a thumbs-up," said Chappell. "Today it's difficult to go out there and pick up something off the floor and not get some form of immediate feedback."

PART IV

ROLES & RESPONSIBILITIES

CHAPTER 12

THE MANY ROLES OF PLAYING IT SAFE: WHO DOES WHAT?

I f behavior-based safety were a play (stage production), the frontline employees would be the actors—out front, onstage, and in the limelight. But actors alone cannot put on a production. The stage must be set with backdrops, props, and all the tools the actors need. There must be lighting to illuminate important areas, appropriate wardrobe, a director to provide guidance and assistance, and all manner of stagehands to ensure the actors have everything they need to play their roles effectively.

As with a stage production where the actors tend to get most of the attention, in BBS the frontline gets a great deal of attention. In truth BBS is a group effort and every role is as important as the next. It is when everyone plays their part well that the production is a success (and has a long run).

The starting point is *pinpointing*. For every position in the company the following question should be asked: what daily and weekly behaviors would improve safety? As mentioned before, the frontline behaviors are those we have called *direct safe behaviors* (they affect the safety of the person performing them). Most other positions in the company will focus more on *indirect safe behaviors* (those that affect the safety of others). Don't confuse indirect with unimportant. The behaviors of supervisors, managers, executives, and safety professionals can make or break the BBS process. Resources must be provided, barriers must be removed, systems and processes must be assessed and reengineered to promote safety, and reinforcement *must* be in place for all those who successfully change their behavior.

Once critical behaviors for all roles have been pinpointed, a positive accountability system will ensure that those behaviors occur.

Here's a look at who does what.

On the Frontline

Frontline people usually like the idea of BBS because the methods are positive and within their control. Some behavior-based safety approaches not only disregard the importance of recognition and reward, but also are inherently punishing by design. Such systems are punishing because they consist of long lists of targeted safe behaviors, require specific observation times that interrupt the workflow, and include feedback sessions in which employees learn about everything they've done wrong relating to the safety behaviors in question. Because so many people associate new initiatives with such negative experiences, it sometimes takes as much as a month of data collection *without punishment* to establish trust in a BBS process such as the one described in this book.

The BBS we recommend is designed to be non-intrusive, easy to do, positively affirming, and therefore, very effective. Observers are not asked to take excessive time from their work schedules, are not asked to observe someone out of their own work group(s), and are never asked to report a peer for noncompliance.

A Summary of How It Works

Each functional group within an organization (for example mechanics, operators or office personnel) form a core team of volunteers comprised of a subset of that work group. This group completes formal BBS training, usually onsite. These core groups from each functional area are responsible for managing the BBS process in their area. This means that they facilitate the selection of pinpointed behaviors, do observations to collect data, provide feedback to the group on progress, and plan and deliver the positive reinforcement. In addition, every person in each and every company work group is invited to complete safety observations and will learn to do so in a short training session provided to all employees. The bulk of the observation task lies with the core team members as well as the responsibility of data collection and updating of feedback graphs. But never fear! It is not as ominous as it may sound.

Suppose your work group is a team of mechanics. A subset of your group has agreed to perform as the core BBS team. With a pocket card listing a few specific behaviors agreed on by the group, the core team members simply work observations into their day when they spot an opportunity. For example, on the way to the workbench to get a tool, an observer may see several team members lifting a piece of heavy equipment. If safe lifting (with backs straight and knees bent) is a scorecard behavior, the observer takes note, makes a tick mark (safe or at-risk) for each person observed (no names included) and proceeds on to the workbench. No big deal! Observers use opportunities like this throughout the day to do quick observations. At the end of the day the group of core team observers has essentially tallied enough observations to constitute a valid random sampling of the group of mechanics. One mechanic then gathers the scorecards, calculates the percentage of safe behaviors for the total behaviors observed and updates a publicly displayed feedback graph or whiteboard for each behavior.

Core team members also take part in recognizing and rewarding people with a quick pat on the back, a wave, or a positive comment. Usually the core teams are given access to a budget for purchasing small tangible reinforcers and paying for celebrations. The entire work group pulls together by discussing the posted data in regular safety meetings and by determining the forms of recognition, reward, and celebrations they would like to receive for success. In other words, the core team members may be the ones organizing the party, but everyone is invited to attend and help.

The Role of Leadership:
Supervisors, Managers, and Executives

The role of all those in a leadership position is critical. Enthusiastic support and effort at the frontline will certainly get the BBS process started, but managers and supervisors who put up roadblocks can bring the effort to a screeching halt. The next chapter provides details on the ways that leaders can influence safety and how to translate those into pinpointed behaviors for supervisors, managers, and executives. Once that is done, establishing mutually agreeable accountabilities for those pinpointed behaviors is the best way to avoid a stalled safety effort.

The exact behaviors executives, managers, and supervisors will engage in vary from group to group and will be established by all parties as BBS is implemented in your area. To ensure these behaviors happen, scorecards are created and an accountability system is established. Executives, managers, and supervisors share their scorecard data with their own managers who also have scorecards for tracking their own BBS accountabilities. Importantly, positive recognition should be built-in at every level and frontline employees should realize that positive recognition and reward works both ways.

One of the biggest opportunities for leaders is ensuring that safety truly is a priority every day. We all figure out what our bosses' priorities are not only by what they tell us, but more importantly by what they pay attention to (what they ask about and what they spend time on). By paying more attention (in terms of antecedents and consequences) to productivity than safety, managers and supervisors send a message that it is better to rush through a job and perhaps take a safety shortcut than it is to do things the safe way. BBS provides managers and supervisors with tools to ensure they talk about safety, ask about safety, and provide positive consequences for safety, just as they do for other priorities such as production or quality.

A great way to assess how well leaders have made this important change is the Safety Leadership Survey. This upward survey asks all employees to answer questions about their immediate boss' safety leadership behaviors. Administration of this quick survey every six months provides supervisors, managers, and executives regular feedback on their safety efforts from the people they support (their direct reports). The feedback can then be turned into action plans for improvement and built into the scorecard and accountability process.

As you will see in the next chapter, executives, managers, and supervisors can and should support safety in many ways.

The Role of the Internal Safety Professional

Most organizations have designated employees as safety coordinators. These individuals may be trained safety professionals or they may be employees who have completed on-the-job safety training. Responsibilities of these safety coordinators include audits, safety training, safety meetings, OSHA compliance, procedure development, and so on. As previously stated, BBS is only part of an effective safety system. All the other pieces of the safety system must be in place and functioning well for BBS to work optimally. The safety professional ensures that these other pieces are functioning and assists with the BBS process. The exact role varies from company to company, but a few standard responsibilities should always be included.

First, the safety professional should assist with coordinating all safety efforts so that BBS is integrated into the system and therefore is sustained. Second, because the safety professional has access to data on injuries and near-misses, he/she should help work groups determine which behaviors to target for improvement and show those groups how to break processes into manageable and observable behaviors. Third, the safety professional should champion the removal of all barriers to safe behavior identified by core teams. Others, such as management, engineering, and maintenance departments may be responsible for actually removing the barriers, but the safety professional should follow through to make sure it happens. The key regarding this role is that the safety professional facilitates the BBS process but is not solely responsible for it. All employees should take a responsible role in the process.

Safety: A Worthwhile and Achievable Group Goal

Everyone in an organization naturally focuses on his/her job duties, getting things done, meeting production goals, and all of the many other responsibilities that seem to conflict with keeping a watchful eye on safety. If the focus on other areas leads to placing safety on the back burner, people will get hurt. People at all levels in the organization must realize that safety and productivity are not mutually exclusive. The problem occurs when an emphasis on one type of performance works to the detriment of another. Safe behavior doesn't have to butt heads with productivity. BBS is designed to allow people to meet all of the demands of their jobs while learning to automatically do the safe thing. In fact, many companies have brought hundreds of behaviors to habit level with a minimal effort that turned out to be fun, meaningful, and an everyday part of the work routine.

Summary

- Everyone has a role to play in BBS.
- Frontline employee groups called *core teams* manage the process for their work areas.
- Supervisors, managers, and executives play a support role to ensure success.
- Safety professionals facilitate the BBS process while continuing to focus on other important aspects of safety.

Did You Know?

- There is no leading cause of death for people who live past the age of 100.

- Emergency room statistics indicate you're most likely to crash into a glass door in the late afternoon.

- If every man, woman, and child in the United States smoked cigarettes, they'd still be outnumbered by the cigarette smokers in China.

- More than half of all the trucks stopped at random in a five-state survey conducted by the National Transportation Safety Board failed a brake test.

- A survivor of a severe brain injury typically faces 5-10 years of intensive services at an estimated cost of over $4 million.

From Richard Hawk and Company, "Safety Stuff" Newsletter. Reprinted by permission of the author. www.makesafetyfun.com

CASE STUDY

Removing Barriers to Achieving Exceptional Safety

Truly effective champions of behavior-based safety (BBS) play many important roles from attaining employee buy-in, to ensuring management support, to sustaining ongoing participation and improvement. But, according to David Andrews, the full-time safety champion for Eurokera North America, one of the most important aspects of the job is keeping the BBS consultant out of trouble. Not to imply by any means that a BBS consultant is an accident waiting to happen—Andrews refers to the key activity of sharing information about navigating the culture of the organization. It is not that Eurokera is particularly political, but all companies have norms or unwritten rules that a new consultant can't immediately pick up on. "I can say, 'If we need to get this done, then we should talk to this person first, get their buy-in, and then we'll go talk to this one.' For example, some people say. 'That's my area; you should have talked to me first!' Certain people you have to have data for; certain people you have to appeal to their emotions. You have to know about an individual's personality so you know how to talk to them to get your point across," Andrews explains.

Andrews is the environmental, health, safety and recycling coordinator for Eurokera North America. Eurokera, established in 1995 and located in Greenville, South Carolina, is the leading glass-ceramic cooktop manufacturer on this continent. The company's 150 employees produce approximately 2.5 million cooktops per year for major appliance manufacturers. A few years ago, the facility decided to involve its employees in a safety effort—to address incident and lost time accidents—by offering bonus incentives for reporting unsafe workplace conditions. The effort included data collection and scorecards, setting a good standard for what was to come. "After we had addressed all of the reported conditions, we noticed that people were reporting behaviors, like someone not wearing their safety glasses. That's when we started looking for a BBS consultant," says Andrews.

A manufacturing process that includes transporting, melting, pressing, cutting and edging, screen printing, heat treatments, and hole drilling of glass carries great potential for a variety of incidents. Eurokera's management team also considered a variety of behavior-based safety specialists, but some of the specialists' approaches were too rigid, and as Andrews saw it, possibly not practical, especially in the area of behavior observations. "Look at it this way," he says. "If a police officer tells you he is going to follow you for the next ten miles to observe your driving, you're going to drive your best aren't you? That doesn't mean you're going to do that when he isn't watching. Also, we didn't want a one-size-fits-all approach. We wanted a process that was tailored to Eurokera, because our culture is very unique."

They found an adaptable BBS approach with Aubrey Daniels International (ADI). Bart Sevin, Ph.D., a behavioral safety consultant with ADI, helped the group design a system that fit the facility's needs while also shaping safe behaviors to habit strength. "David is an extraordinary site champion," says Sevin. "He knows everyone in the plant well, he is trusted and respected by people at all levels, and he was instrumental in helping me effectively negotiate the politics of the plant. His guidance helped me avoid more than a few pitfalls and greatly facilitated the successful implementation of the BBS process."

"Bart trained everyone in the behavior basics, so that everyone knew what to expect and didn't panic when people started writing stuff down," says Andrews of the peer-to-peer behavioral safety observations. Eurokera adopted a simple process in which peers take a few minutes to look over at a co-worker to see if he or she is engaging in a pinpointed behavior such as using personal protective equipment. They then mark a checklist and provide real-time feedback. Each group also tracks and charts its own data. Importantly, the primary focus is on observations and feedback, rather than on lagging indicators such as reduced incident rate. The entire plant participates in the safety initiative with the same celebration budget allotted for every level of the organization. After a group meets a subgoal or goal, a short meeting recounting the reason for the gathering and thanking everyone for their efforts precedes a celebration designed by the participants. "Probably one of the best things we've done is keeping the celebrations simple," says Andrews. "We don't have the escalating celebrations that I've seen turn into nightmares for numerous companies."

Andrews shows up for these events even in the wee hours of the evening, often stopping by an all-night restaurant to pick up the food that the late shift has ordered for its celebratory meal. This effort is greatly appreciated by those who know Andrews works during the day; yet he considers their achieved safety milestones valuable. "They figure if the celebration is important enough for me to get out of bed at one o'clock in the morning and come in to meet with them, then the safety observation cards are important enough for them to do," Andrews comments.

The three years that Andrews has taken on the role of internal safety champion for this Eurokera facility have provided him with lessons learned that he shares in presentations to other facilities—lessons not only for designated safety champions but for safety leaders. According to Andrews, a safety leader can be anyone who is willing to continually encourage and reinforce others to be proactively safe. As he puts it, "the devil is in the details." For example, good safety leaders ask general questions about BBS such as "How is BBS going?" or "Are you doing observations?" However, exceptional leaders ask specific questions: "What did you do to improve BBS today?" "Who did you talk to and what did you say?" "What have you done to hold those who report to you accountable for performing their role in BBS?"

One of those exceptional leaders has been Bill Mountain, former plant manager. "He embraced BBS as plant manager and when he was promoted to president continued to be active in the process," Andrews explains. "He filled out a card every day. He went out on the floor and conducted safety observations. His ongoing participation showed that he was behind it."

Andrews also readily lists the "high-impact" behaviors of an internal safety champion, a few of which are as follows:

- Frequently contact and coach core teams, managers, and supervisors

- Directly and indirectly facilitate the effective use of antecedents and consequences with people at all levels (accountability)

- Track outcomes and success measures, and give frequent updates to people at all levels

- Ensure consistency in all aspects of the process

- Proactively bring potential issues and problems to leadership

Andrews categorizes "keeping the consultant out of trouble by introducing him to the inner workings of the culture" under the high-impact behavior of "facilitating barrier removal." This means removing barriers that might impede both the implementation of BBS and the achievement of safety goals. Therefore, in addition to sharing information with the consultant about the internal culture, he may facilitate the relocation of an earplug station to help a team achieve the behavioral pinpoint of wearing earplugs at habit strength.

"I don't have disciplinary authority. I can't say to do this or else," says Andrews. "I have to help sustain this process by finding a way of interacting with people that clicks with them about how important this is." For example, when Andrews shares data as to the impact of BBS, he not only discusses the numbers, but personalizes the information with details such as, "This is how many people got stitches this year as opposed to last year."

"You have to be able to show people that it's working!" he asserts.

The data is progressively easier to share. Since the day Eurokera's employees started the process three years ago, as the participation in conducting safety observations rose, the recordables and lost time accidents diminished. When the facility began the BBS process, the recordables stood at 24.2 and the lost time accidents at 2.69. Currently 97 percent of the frontline employees conduct peer-to-peer observations and feedback. With a recordable rate at 4.2 and lost time accidents at zero, Andrews doesn't think the improved safety record is a coincidence. He concludes, "We've made tremendous progress with BBS."

CHAPTER 13

SAFETY LEADERSHIP: A PROACTIVE APPROACH

SAFETY IS OUR FIRST PRIORITY

Safety is our first priority. Making this slogan a reality is a good way to sum up the role of leaders in BBS. While most supervisors, managers, and executives believe that the safety of the people who work for them is truly the top priority, making that a reality is more difficult than it seems. Because most organizations exist to make money, the focus easily shifts to those things that create revenue. When no one is getting hurt, it is easy to ignore safety and focus on productivity, quality, customer service, and so on. Thus, many companies find themselves managing safety in a largely reactive way—reacting to accidents, incidents, and near misses when they happen but shifting back to focus on production once the incident seems to be "under control." The way we measure safety contributes to this less-than desirable approach. When the only measures of safety are results—such as incident rate—then falling into a reactive management approach is natural.

BBS provides a solution to this dilemma and allows all leaders (supervisors, managers, and executives) to identify leadership behaviors that prevent accidents and thus manage safety proactively.

We will continue to describe the behaviors that leaders engage in around safety as *indirect safe behaviors.* They are *safe* behaviors in that they promote safety, but they are indirect in that they typically affect the safety of others (frontline).

In this chapter we will discuss two categories of behaviors that leaders should focus on and how to use the behavior-change strategies outlined in this book to ensure those behaviors happen.

Two Ways for Leaders to Influence Safety

Supervisors, managers, and executives influence safety in two general ways:

1. Direct application of antecedents and consequences for safe and at-risk behaviors.
2. The creation and maintenance of systems and processes that impact safety.

CASCADED ABCs

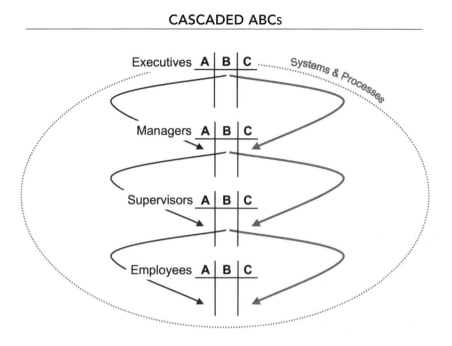

Direct Application of Antecedents and Consequences

Perhaps the most obvious way that leaders can influence safety is by providing antecedents and consequences for safe behaviors. As the diagram shows, this is done in a cascading manner. Supervisors (or team leaders) prompt and reinforce frontline safe behaviors. Managers prompt and reinforce supervisors for indirect safe behaviors (safety leadership behaviors). Senior managers prompt and reinforce managers for indirect safe behaviors, and so on.

This cascading flow is a primary source of consequences. However, anyone at any level can reinforce anyone else, provided they know what the pinpointed safe behaviors are. The goal is to increase the amount of positive reinforcement all employees receive for safe behaviors (direct or indirect). Given that in many companies the bulk of safety consequences are negative, there must be enough positives to tip the balance so that safe behaviors fall into the *want to do* category rather than the *have to do* category.

Keep in mind that it is not just intentional antecedents and consequences that are impacting safe and at-risk behavior. In safety, many unintended antecedents and consequences operate. Therefore, one important role of management is to learn to analyze the impact of their actions to avoid unintentionally prompting and reinforcing at-risk behavior.

Here is a common example. Frontline employees sometimes speed while operating equipment. A PIC/NIC Analysis shows the PICs that exist for speeding.

PIC/NIC ANALYSIS

Operator At-Risk Behavior:
Speeding while operating equipment

EXAMPLE #1

CONSEQUENCES	P/N	I/F	C/U
Get work done faster	P	I	C
More fun/exciting	P	I	C
Stay on schedule	P	I	C
Praise from boss for productivity	P	F	C
Get hurt	N	I	U

As you can see by this analysis, there are three PICs and one PFC for speeding. There is only one negative and it is uncertain, therefore weaker. Closer inspection shows that three of the positive consequences for unsafe behavior are related to better productivity. This tells us that there are probably many intentional and unintentional ways that working quickly (and therefore getting more work done— or perceiving that you are) is reinforced in this workplace.

Does that mean the supervisor is to blame? No. Let's do a PIC/NIC Analysis on supervisor behavior. A common at-risk behavior that supervisors engage in is praising workers for high productivity (or meeting deadline) without checking to make sure the work was done without taking safety shortcuts.

PIC/NIC ANALYSIS

Supervisor At-Risk Behavior:
Praising operators for productivity without asking about safety

EXAMPLE #2

CONSEQUENCES	P/N	I/F	C/U
Feel like I am managing productivity	P	I	C
Easier to manage by results	P	I	C
Better productivity/ stay on schedule	P	F	C
Praise from my boss	P	F	C
Avoid conflict about safety	P	I	U
Operator gets hurt	N	F	U

In this analysis, like the operator analysis, there are more positive consequences than negative for making safety communications secondary and the positives are more powerful than the negative. The supervisor (like the operator) is choosing the at-risk behavior because the consequences support doing it. It is all too easy for leaders to have good intentions regarding safety, but to unintentionally have a negative impact.

A PIC/NIC Analysis could be (and should be) completed for all levels of the organization. This highlights how people and systems unintentionally promote and support unsafe behavior and points toward possible solutions. As mentioned before, one piece of the solution is creating frequent measures of safety so that leaders can provide PICs for safe behavior each day just as they provide PICs for productivity. Another piece of the solution is discussed below.

Creating and Maintaining Systems That Impact Safety

"Our experience has led us to a bias: most people want to do a good job. However, if you pit a good performer against a bad system, the system will win almost every time."[1] This quote by behavioral researchers Geary Rummler and Alan Brache points out the importance of looking at the systems and processes in the work environment as sources of consequences. Heaps of sincere positive reinforcement from a supervisor will not offset the fact that short-cutting a safety procedure, and thereby getting more work done, will result

1 Rummler, G.A. and Brache, A.P. (1995). *Improving Performance: How to Manage the White Space in the Organization Chart.* San Francisco: Jossey-Bass, Inc.

in a large bonus. Similarly, recognition from peers is unlikely to offset the fact that old equipment makes working safely very difficult. These are examples of barriers to working safely. Many of the barriers identified while implementing BBS are based in systems and processes created and maintained by management.

Examples include
- work procedures,
- incentive systems,
- measurement systems,
- quality procedures,
- equipment,
- staffing,
- supplier/contractor relationships.

All of these have potential implications for safety. Thus, an important task for management is to assess existing systems and their impact on safety.

A related task for management involves considering safety in all decisions. We recommend that all leaders work on asking the following question *habitually*: "how will this decision impact safety?" In other words, before any decision is made, the potential impact on safety should be assessed. This sounds simple, but to answer this question effectively, leaders must understand behavior from a scientific perspective. They must realistically look at the antecedent and consequence changes that each decision will make (intended and unintended) and predict the impact on safe and at-risk behavior. For example, if a new measurement scorecard is implemented across the company to track important business objectives, where is safety on the scorecard? What weighting does safety have? If incentives or bonuses are tied to the scorecard, what behaviors will that encourage and discourage? Keep in mind, every management decision changes behavior and every behavior change potentially has safety implications.

A dramatic example comes from a railroad company that instituted a Six Sigma initiative aimed at turning rail cars faster (uncoupling/recoupling cars and engines). It took six months to plan and execute the change and the projected savings were $1 million annually. Within the first few months of implementation an accident occurred that caused untold human suffering and cost the company $6 million. In hindsight, the accident was predictable but only with a scientific understanding of behavior.

> There is no decision, no process, no policy, no procedure, no change that should be made without considering safety. After all, "Safety is our first priority."

Summary

Following are the general categories of behaviors required at each level.

SUPERVISORS

- Prompt and reinforce safe behaviors of direct reports (frontline).
- Provide constructive feedback regarding at-risk behaviors.
- Identify and remove barriers to safe behavior.

MANAGERS

- Prompt and reinforce indirect safe behaviors of direct reports (supervisors).
- Remove barriers to safe behavior.
- Prompt and reinforce safe behaviors of frontline associates.
- Assess the impact of existing systems on safety.
- Ask "how will this impact safety?" when making any and all decisions.

SENIOR MANAGERS/EXECUTIVES

- Prompt and reinforce indirect safe behaviors of direct reports (managers).
- Remove barriers to safe behavior.
- Reinforce safe behaviors at all levels.
- Assess the impact of existing systems on safety.
- Ask "how will this impact safety?" when making any and all decisions.
- Provide resources for safety.

Did You Know?

- Marie and Pierre Curie's notebooks recording their radium experiments are still radioactive.

- When people are wide awake, alert, and mentally active they are still only aware of 25 percent of what various parts of their bodies are doing.

- 100 Billion: It's the estimated annual loss in dollars because of sleep-deprivation related problems in the United States.

CHAPTER 14

SELF-OBSERVATION:
TRUTH AND CONSEQUENCES

"We had some real hard-core loggers in our behavior-based safety training and we knew their buy-in was essential to the success of the process," said John Venasky, safety coordinator for Avenor's Wood Products Group in Canada. Individual buy-in was even more essential at Avenor because the working logistics required self-observation, and forestry industry loggers and equipment operators work anywhere from a hundred feet to several miles apart.

"It's pretty difficult for a guy who's cutting wood to stop, get off a tree feller/buncher machine, and drive several miles to observe somebody else operate their machine," stated one forestry employee.

Lone workers or *remote workers,* those who work at a considerable distance from others, must be exceptionally careful. Response time to an accident in an isolated area is almost inevitably delayed, so these workers are at higher risks should a serious accident occur. But how can an individual worker, caught up in daily duties, monitor his/her own safety on a regular basis? Workers on deep-sea oil riggers, loggers virtually alone in forested areas, long-distance drivers, railroad repairmen, and yes, even remote office workers have successfully used a behavior-based, self-observation method to ensure their own safety.

Usually, self-observers (like all BBS observers) use a simple card method to observe their own behaviors and collect data on those behaviors several times a day. The self-observer has a scorecard, just like in peer observation and simply checks *safe* or *at-risk* (depending on whether or not he/she was performing that behavior safely at specific times set for self-observation). Lone workers then share this data with a peer data collector who graphs group progress just like the peer observation approach. As one might imagine, self-observation brings up several unique issues.

The first issue is "Why wouldn't people just mark all *safe* and be done with it?" Well ask yourself another question, "Are the majority of people who work in your organization honest?" Most people answer that question with a yes. This brings us to ask, "What would make a normally honest person do something dishonest, or essentially, lie? Why would a person put all *safes* down knowing that doing so could jeopardize his or her personal safety?"

> The only reason a usually honest person would lie is because of the perceived consequences for doing so.

People may balk at collecting honest data on their own behavior for fear of negative reprisal. If they put a tick mark under *at-risk*, they may be punished. An immediate and frequent response upon hearing about self-observation scorecards is, "Of course, I'll get reprimanded. It's self-observation and my scorecard shows I have done some at-risk behaviors!"

BBS methods make sure that individual self-observation data is not punished. Names are not written on the scorecards and individuals turn the scorecards in to a peer who collects the group data and then destroys the card. This way, even if a manager happens to see a scorecard, the data cannot be attached to a single performer. With BBS, managers receive extensive coaching on the importance of not punishing the at-risk but rather focusing on and positively rewarding improvement. With lone workers it is even more crucial that the behaviors chosen for monitoring truly have a meaningful impact on safe performance.

Another perceived consequence that might encourage someone to mark all *safes* on a self-observation card is if the performer thinks or knows a large tangible reward is available. If lone workers were told they would receive $1000 per person for turning a particular behavior into a safe habit within a month, then cheating is a strong possibility. Therefore, BBS methods discourage such large rewards, instead stressing the importance and impact of immediate, small, frequent rewards and reinforcers.

Another important, though long-term, reward exists. If you're working alone, you are the only person who can keep you safe. So if you cheat on self-observation data, the only person who gets hurt is going to be you. Most people figure this out pretty quickly.

Triggers to Self-Observation

Even if you are an honest person, support the self-observation process, and want to adhere diligently to safety standards, it goes against human nature to choose to observe oneself in the middle of doing something wrong. Chances are, if you've developed an at-risk habit, you won't even notice when you are in the midst of it.

The trick to solving this problem is to set up triggers/reminders (essentially antecedents that tell you to do a self-observation) that occur periodically through the day. For example, you might set your watch alarm to go off at regular intervals to cue you to observe what you are doing at that moment. Law requires most companies to schedule regular radio/telephone/beeper check-ins with remote workers involved in risky work. Those check-ins would be an ideal cue for self-observation on performance of the desired safe behavior(s). For example, truck drivers may ask dispatch to call periodically during the day with the reminder to check at that moment the specified behavior(s) as in safe following distance, correct speed limit, and so on. Most people can think of triggers to self-observation that are very specific to their everyday work duties. Office workers, for example, may perform a self-observation on a chosen safe behavior whenever their computer signals that a new e-mail has arrived. Some people check ergonomic behaviors such as a pinpointed correct sitting posture and mark their observation cards whenever the phone rings. Long-range drivers sometimes hang a reminder on the rearview mirror or place a sticker near the speedometer to prompt them to regularly check a behavior.

If performers have a difficult time remembering to do certain safe behaviors, they may use other prompts such as a sticker placed on the receiver of the phone as a reminder to hold the receiver properly rather than cradling it between head and shoulder. If the prompt does its job and reminds the person to perform the task the safe way, then that person can mark a *safe* on the scorecard.

Self-Observation and Self-Reinforcement

Essentially lone workers must do a lot of on-the-spot self-recognition for making safety improvements on the job. However, organizations can reward and reinforce group improvement as well as individual improvement (if and only if individuals feel comfortable in sharing personal data) with lone and remote workers just as they do with teams that work in closer proximity to one another. In addition, if the data collector (person who collects and summarizes the data collection cards) has a good relationship with peers (as he or she should) that person can offer positive remarks when the cards are turned in. Also, remote workers can find ways to reinforce and recognize one another as a group by getting together for meetings, sharing data and encouragement through e-mail, radio contact, or in group celebrations for reaching goals and subgoals.

The good news about self-observation is that at the end of the day when you look at your scorecard and see there are more tick marks under the safe column than under the at-risk column, you receive immediate feedback. If presumably you really want to improve in that behavior then it is reinforcing to see those tick marks shift from the negative side to the positive side knowing that it signifies you are safer today than you were the day before.

The hard-core loggers of the Canadian forestry company mentioned earlier were surprised that this was indeed the case once they agreed to give self-observation a try. This group of rugged individualists found that learning was doing and quickly became proponents and participants of using the self-observation card for improving safety. The organization's safety coordinator summarized it this way:

"I think most employees get a real boost because this card is a contract they've made with themselves. I have guys tell me that their objective is to have a clean card of *safes* for the week. They work hard and get disappointed when they have to mark even one.

"I equate it to a golf card. You have to be honest with yourself. When you score a six on a hole, you have to mark a six or you're only lying to yourself. When they see they're doing well on the card, they're actually patting themselves on the back. They get that real sense of accomplishment.

"I don't think it matters what walk of life you're in. At the end of the day, when you know you've actually accomplished something, you have that good feeling that you've contributed. Because they are measuring themselves, they now have more opportunities to look and say, 'Yes, I *did* contribute.'"

Summary

- Lone workers can participate in BBS through self-observation.

- Because the self-observation process protects anonymity and ensures that only positive consequences are used, self-observers quickly become comfortable doing honest observations of their own behavior.

- Accurate self-observations require catching yourself acting naturally, so prompts are needed to trigger observations periodically during the day.

- Self-observation has worked effectively for a wide variety of lone workers in many industries.

Did You Know?

- Most accidents that befall air travelers occur on the ground rather than in the air: tripping over luggage, taking a spill while running to the terminal, etc.

- Three million people in the U.S. have impaired backs or limbs as a result of an accidental fall.

- Fewer than 1 percent of people over seventy-two years old are able to walk fast enough—2.7 miles per hour—to make it across an average intersection before the light changes.

- You are 1.6 times more likely to get killed by a car while out for a stroll than you are to be shot and killed by a stranger.

From Richard Hawk and Company, "Safety Stuff" Newsletter. Reprinted by permission of the author. www.makesafetyfun.com

CASE STUDY

A Culture of Safety Ownership at Malt-O-Meal

Malt-O-Meal, headquartered in Minneapolis, Minnesota, began an American culinary tradition when it introduced its hot cereals in 1919. In 1965, the company manufactured its first ready-to-eat cereals and has since become the fastest growing cereal company in the country. Rapid growth—today Malt-O-Meal operates six plants and eight distribution centers—hasn't compromised the company's dedication to product safety or to the safety of the employees who produce those products.

In 2003, Mark Suchan joined Malt-O-Meal as the plant manager of the company's first facility outside of Minnesota in Tremonton, Utah. "For Malt-O-Meal it was quite a learning experience and I was given a lot of latitude in some of the things we tried as we started up the plant," says Suchan. One of the first new things that Suchan tried was a behavior-based safety (BBS) effort. The company didn't have a behavior-based process in place, but Suchan was familiar with the concepts and attempted to get, what he describes, as an "unstructured" process going. "It was really kind of a ragtag program," he admits. "We didn't have all of the resources we needed to get it off the ground and to support it."

But Suchan wasn't deterred. He met with Bob Johnston, the corporate director of safety. "Bob has been a big proponent of BBS. I said, 'I'm just not getting the traction. We just don't seem to have the culture of safety. We talk about it and do some of the right things, but we don't see the results and we don't have real buy-in." At this point the facility was experiencing an incident rate above industry average. According to Suchan, Johnston went on a quest to find a system that he felt would work. He found Aubrey Daniels International (ADI). "Its process appeared to match our company philosophy and seemed capable of taking us where we were trying to go. We wanted to ensure we had a program that could be applied against any system we were trying to improve. It needed to provide consistent reinforcement for the right behavior for any process we were trying to get to habit strength," Suchan adds.

Feeding the Right Culture

Malt-O-Meal wants to nurture a culture in which employees are accustomed to autonomy and the ability to make decisions. "We want our operators to feel ownership in our processes all the way through: quality, food safety, and personal safety," Suchan explains. "We wanted them to have that ownership but some were just hesitant to take it on. We just weren't creating that culture of safety and personal ownership of safety."

When Bart Sevin, Ph.D., an ADI behavioral consultant, trained employees in peer-to-peer safety observations, neither they nor Suchan realized that it was the first key to creating a culture of team ownership. "At first we had to learn to approach each other and give each other feedback," Suchan says, "That took a bit of learning." That first step broke the ice and when employees learned that joining a safety team was voluntary, their participation began to grow. However, the management team faced a different type of challenge: they had to let go.

"For the management team, the difficult part was getting out of the way," says Suchan. Did the management need to micromanage the pinpoints the teams chose? Did they need to interfere with how the teams spent their celebration budgets? Was it necessary for management to determine which team members performed observations and when? For example, Suchan relates that early in the process management complicated matters by collecting all of the data and giving one person the role of compiling it into meaningful feedback. It just didn't work. The lesson learned: keep it simple and let the team members run the process!

"The temptation as managers is to say, 'Do it this way; spend your money that way; abide strictly by this rule.' It was tough for us not to do that, but once we really said that this was their program, let them define it, let them figure out how to run it, once we got over that hurdle, BBS took off and the flywheel was just spinning," Suchan says.

Team Ownership—the Safest Alternative

After a year and a half, the Tremonton facility applies BBS at all levels of the organization with 25 safety teams consisting of anywhere from six to ten people, with two to three core team members from each group. Even though Suchan occasionally suggests a behavioral pinpoint, the teams select their own pinpoints—up to three behaviors per safety checklist. They also collect data and chart their own progress for pinpoints such as washing hands using a defined process, wearing hair nets properly, donning gloves, or using ear plugs.

With a goal of reaching habit strength for each behavior—defined as everyone doing the safe behavior all of the time—the teams repeatedly demonstrate that not only can they reach those goals, they understand the importance of admitting when they haven't. For example, team members may be more than halfway through to reaching a celebratory milestone when someone fails to complete a pinpointed behavior. Invariably the team reports the stall and starts again from the beginning. "We regularly compliment them on their integrity and encourage them to keep up the good work," says Suchan.

Celebrating Success

"We do a lot of celebrating," Suchan states. A lot of celebrating indicates a lot of success. Teams set their own subgoals and determine whether to spend their budgets on a series of small celebrations or on a larger reward once a behavior reaches habit strength. Often the team will celebrate subgoals with a purely social celebration or by bringing in small treats for the team and saving the budget bucks for a bigger event. Yet, they don't carry funds for longer than a month after reaching a goal in an effort to adhere to the immediacy rule of reinforcement. Either way, the fact that they design their own reward is reinforcing in and of itself. "It's a big deal to give the teams control over the celebrations," Suchan acknowledges. "They have a lot of fun with it and they come up with some pretty creative stuff. As for the process, it's got momentum now for sure."

Sustaining Safety

Part of that momentum depends on reinforcement from management The management team makes a concerted effort to reinforce the process by going out on the floor and asking people how they are doing. Once a week they meet with the core team members who tell them about their pinpoints and share their progress.

Sevin, the BBS consultant, arrives quarterly for "tune-up meetings" offering advice and tweaks to the ongoing safety efforts. At one time, only core team members were invited to these learning sessions but now that everybody is performing observations, the meetings are open to anyone who wishes to attend. Regular team meetings include discussions of safety topics, a quality topic, and any operational issues for the day.

The Right Mix

The Tremonton facility made a good decision early on to designate three site champions—the operations line manager, the quality manager, and the safety manager. This combination has proved successful because it didn't burden any one person with too broad of a scope in such a large organization. "Their day jobs, if you will, are a natural fit," Suchan explains. "We will always want to improve quality, we always want to improve food safety, and we can always improve on personal safety. We want to make sure that everyone goes home safe and those are issues that are easy to rally a team around. But we've also seen that we can use this approach whenever we have a new policy or procedure, a new way to measure ingredient yields, whatever the change may be; this is just a good way to implement change period."

In less than two years the employees at Malt-O-Meal's Tremonton plant have learned that autonomy can extend to teams without limiting that of the individual. They've grown to enjoy the power they have to bring about positive change and the impact they have on achieving significant company goals not necessarily limited to safety. Suchan believes they have also achieved a culture of ownership for their own safety as well as that of their teammates. That is reason enough to celebrate and the fact that the plant has already reduced its incident rate by half is icing on the cake!

PART V
THINGS TO CONSIDER

TAKING SAFETY HOME

Phil Gardner, a superintendent of technical services with a Canadian forestry products company decided to test the BBS methods used successfully by his organization on his own personal habits. He knew that he should use his safety belt every time he got behind the wheel, but he frequently failed to do so. It took a few seconds and a simple scorecard marked with a yes or no category to develop a lifelong habit that will possibly save his life. "Every time I got in my truck, I'd either put my seat belt on or I wouldn't," he said. "The moment I realized I had or had not put on the belt, I'd take out the card and check a yes or a no. The no answers were winning and it upset me. I kept plugging along, with the yeses gradually increasing, until after maybe two or three weeks there were no more no answers. It was done!"

Gardner continued to wear his seat belt and to use the card method. Some of his coworkers decided to do the same, targeting personal habits for change. One associate used the self-monitoring card method to stop himself from regularly using profanity. Another successfully used the method to cut down on soft drink consumption. Yet another used the card to become a better listener when his wife was speaking to him.

"We discovered very early with this behavioral stuff that until you're consciously aware that you're doing an unacceptable behavior, you can't make a change. Experiencing the value of the process to make personal changes also demonstrates that the same method can be used on a work habit. People who use it, when they make a change, are real believers in the system," said Roger Shott, a forestry industry brush saw operator.

Most behavior-based safety professionals will tell you that these kinds of stories are most rewarding to hear. Realistically we know that things change at work and that despite our best intentions, some organizations do not follow through in their support of BBS. However, if someone takes the process home, we know we've changed something for the better. The BBS methodology is based in the science of applied behavior technology, so its tools—observation, pinpointing, measurement, feedback and reinforcement— can be used to impact non-safety-related behaviors as well. (More about this in the next chapter.) Some organizations, such as the forestry organization mentioned above, actually requested that employees try out the self-feedback card approach applied to a personal problem *before* they tried it with a safety-related behavior. However, usually the opposite is the case. Time after time, people who have experienced the ease and success of BBS at work have then brought the process home to deal with a variety of troublesome habits.

Charlie Venishel, a BBS coach with Rochester Gas & Electric (RG&E) caught the positive recognition bug after his exposure to BBS at work. He soon began getting a kick out of the surprised expressions on the faces of store clerks after he made a positive comment about their service. And he saved himself some bucks by promoting healthy habits in his teenage son:

"I put thousands of dollars in his mouth with braces but it was a real chore to get him to brush his teeth regularly," said Venishel. He explained the concept of habit level to his son and let him choose a reward for developing habit level on the tooth brushing and several other behaviors. Even Venishel was surprised at how quickly his son set up his own graph and achieved the goals.

Craig See, an RG&E division safety coordinator commented, "Some of the toughest guys to convert to BBS methods were the ones who said, 'You know, I tried this with my kids and it works. Maybe there is something to this!'" In fact, See tried it himself when he realized that the negative approach to his daughter's lack of commitment in college wasn't working. Realizing she wasn't studying and that her grades were suffering, See chastised her with fatherly statements such as, "I'm paying for this!" and "You gotta buckle down!" When her first semester grades limped in, See changed his tactics. "She didn't do well the first semester, but the second semester when she called, I started reinforcing small improvements," he said. "If she said she was studying, I'd say, 'That's great!' The next semester she did exceptionally well and, in fact, earned a certificate of merit. It was certainly nothing I did except to reinforce what she was doing herself."

Of course these people and many more like them learned that behavior-based safety methods learned at work apply to other areas of life. However, just as many people have reported applying the methods to being safer at home. BBS users report taking the methods home to make habits out of such behaviors as wearing protective glasses while mowing, observing their own behaviors when using gardening equipment and indoor utilities, and monitoring and giving feedback on the unsafe and safe behaviors of their family members. One woman even used the methods on the way home. She pinpointed the beginning behavioral signals of her own rising road rage, countered them with incompatible behaviors (behaviors that are opposite to rage and would prevent her from getting mad, for example doing relaxation exercises) and succeeded in managing and stopping her anger outbursts before they began.

Taking these methods home is no small accomplishment considering that so many accidents occur near or in the home where we usually consider ourselves to be the safest. Just imagine the numerous behaviors we engage in away from work and the potential for injury—driving, walking amidst parking lot traffic, crossing the street, grocery shopping, yard work, cooking, cleaning, lifting, sports, even dining out. One author of this book (whose name will go unmentioned, but whose initials are G. S.) set the entire table on fire at a fine restaurant when her menu dipped too close to a candle!

Our nation's roads and highways alone offer numerous personal opportunities to alter our behavior with BBS interventions. The U.S. Department of Transportation reports that automobile crashes are the leading cause of death for persons between the ages of 6 to 33 years. Drivers of company-owned vehicles have successfully monitored themselves on behaviors such as driving at a safe distance, making full stops at traffic signs and lights, and traveling at safe speeds. Once these behaviors reach habit level at work, they remain consistent for the driver in all driving venues.

Mind-boggling traffic and the rising occurrence of road rage events are reason enough for us to consider using behavior-based safety strategies in our daily private lives. Yet, at-home and near-home activities present even more opportunities for accident and injury. Surprisingly, metal cans injure more Americans each year than the combined injuries of mopeds, all-terrain vehicles, chain saws, and minibikes. On average, a pedestrian is killed in a traffic accident every 96 minutes. Nationwide, bicycle deaths outnumber plane deaths by four to one. Of course, home-related accidents and injuries have no geographical boundaries as evidenced by the following information from the United Kingdom:

- Over 2.8 million people go to hospitals each year as a result of an accident in the home.
- Cushions and pillows cause 3000 accidents a year and beanbags cause an estimated 828 accidents annually.

- In one year, 39 accidents were caused by tea cozies.
- The most dangerous piece of household furniture can be the bed–causing over 100,000 injuries every year.
- Tables and chairs are involved in an estimated 130,000 injuries and a further 1675 by deckchairs alone.
- Over 3500 accidents annually were caused by tripping over laundry.
- 9000 accidents were reported involving vacuum cleaners.
- Accidents in the bathroom were common with an estimated 28,000 involving the bath and 11,000 involving the toilet.
- Cotton balls caused more accidents than razor blades with a whopping 9000 a year.
- Sponges and loofahs caused 630 accidents and talcum powder accounted for over 200 incidents.
- In the kitchen, knives caused over 38,000 injuries, forks and spoons a further 2000.
- And, over 1300 incidents involved baking tins getting stuck on heads!

If these sometimes-bizarre statistics tell us anything, it is that wherever a human being exists, there is potential for accident and injury. Of course, we cannot live our lives in fear, but we can consciously and actively attend to the behaviors that we control that will better ensure our safety. So, challenge yourself! Ask your children to use a card to monitor their use of protective sports equipment such as bicycle helmets and in-line skating protective gear, for example. Examine your own behaviors and that of your loved ones and share with them this system for being as safe as possible and staying that way at work, at home, and in the world at large.

Summary

- BBS methods can be used in other areas of performance but they can also be used to keep you safe on your journey to and from work, at home, and in every moment of leisure.

- For many, trying BBS methods first to change a personal habit is a good approach to transform skeptics into BBS believers.

Did You Know?

- In the 1990s enough people were saved by their seat belts to fill Yankee Stadium to capacity.

- More people are bitten by other people on Saturday afternoon between 3 p.m. and 5 p.m. than any other time of the week.

- Over 1,500,000 Americans are injured each year while sliding into base during a softball game.

- Each year over 100,000 American children end up in the emergency room because of their toys.

- Barnyard hogs kill more people every year than sharks do.

- About 3000 people sought emergency medical attention in 1997 for scalp lacerations, corneal abrasions, ankle sprains, bruises and cuts from clothespins and clotheslines.

- Every 45 seconds a house catches on fire in the United States.

- In Europe, designated drivers place their car keys in their glasses to indicate to a host that they should not be served alcohol.

From Richard Hawk and Company, "Safety Stuff" Newsletter.
Reprinted by permission of the author. www.makesafetyfun.com

CHAPTER 16

BEYOND SAFETY

By now you can see that behavior-based technology can be applied to areas other than safety. At home it can be applied to coaching sports, helping kids do better in school, and improving personal health and fitness. At work, it can be used to improve any business objective. The steps are the same: pinpoint the desired result, pinpoint the behaviors that will lead to that result, measure progress, provide feedback to all performers, and positively reinforce improvement.

The case studies that follow provide a sample of some of the astonishing outcomes this technology has helped organizations achieve. The industry and area of improvement are listed below for each case study.

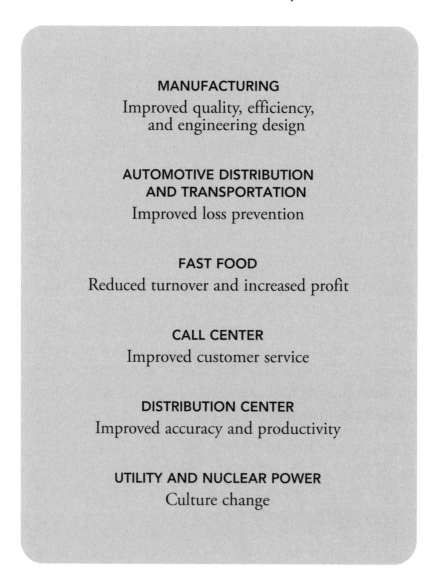

MANUFACTURING
Improved quality, efficiency,
and engineering design

**AUTOMOTIVE DISTRIBUTION
AND TRANSPORTATION**
Improved loss prevention

FAST FOOD
Reduced turnover and increased profit

CALL CENTER
Improved customer service

DISTRIBUTION CENTER
Improved accuracy and productivity

UTILITY AND NUCLEAR POWER
Culture change

CASE 1

High Technology Manufacturing: ROI through Quality, Efficiency, and Engineering Design

With a strong emphasis on continued education, this manufacturer of high optical and high strength glass offers employee training opportunities ranging from hands-on production to associate degrees and management education. When management and supervisory staff completed training in behavior-based technology, they realized they could use it to reinforce participation in the continued education programs. From there, the behavior-based approach expanded to become the driver for all initiatives from production, to engineering design, to quality, to safety. The results of this implementation were soon directly linked to big savings and a significant rise in gross profits.

Situation

As one of the world's primary producers of television face panels, high strength glass, and widescreen and high definition television (HDTV) products, this 50-year-old Japanese-owned manufacturing facility leads the specialty glass industry. However, the demands of producing over 10 million flawless glass units per year, plus the continued evolution of the technology, forced the executive team to ask "How do we retain and expand our market leadership in the television glass industry while continuing to provide a high quality product at competitive prices?" Competing in a fast-paced high tech industry was difficult enough, but the facility also had decades of negative management history to overcome with its largely union employee staff.

Solution Implemented

The management team adopted the positive, employee-driven behavior-based approach to revamp the gainsharing system, structuring it to be contingent on employee behaviors that brought return on investment to the company. Performance improvement teams focused on interdepartmental communications, process redesign, operator discretionary effort, safety, quality and productivity. The facility also developed a computerized performance data collection system for operators paired with graphed feedback and ongoing recognition and celebration. The design engineering group alone attributed a cost savings of $58,000 per quarter to the use of behavior-based performance improvement initiatives. Long-standing hard feelings between work groups dissipated and employees began voluntarily using the methods to improve customer relations and supplier performance.

Results of Intervention

- Produced above-average payouts through gainsharing program based on employee behavior

- Achieved record productivity goals with the same number of employees. (With a focus on rewarding operator discretionary effort, machine utilization efficiency rose from 65% to

85.6% within months. 16 new productivity and efficiency records were set.)

- Lowered injury rate with behavior-based safety process and saved money by targeting responsible use of safety equipment, resulting in 9.5 cents saved per man hour worked

- Realized a net gain of $230,000 per year through improvement of *one* operational inventory change

- Saved $58,000 per quarter from behavior-based initiatives by the design engineering group

- Reduced labor grievances by 30%

- Experienced 8% overall productivity improvement; increased overall plant throughput by 11%; raised good customer service measure by 14%; cut forming defect losses by 50%

- Reduced outside construction costs by 50%

"Previously unheard of levels of improvement over prior periods have occurred and played a prominent role in the plants ability to supply product to customers in a tight market."

– CONTROLLER

CASE 2

Automotive Distribution and Transport: Product Damage Prevention

The largest automotive distributor in North America used behavior-based methods to cut product damage claims by half within one year (a $4 million dollar savings), reduce driver turnover to an unprecedented level, and win quality honors from Ford Motor Company, Toyota, and the National Automobile Transporters Association (NATA).

Situation

This thriving automobile transporter wanted to address typical annual damage claims in excess of $8 million. The smallest nick or scratch on the new automobiles they delivered could be highly expensive. Occasionally damages did occur due to loading, tying and unloading errors, but the carrier was also liable for repairing any pre-existing damage not noted during the pre-loading inspection. Despite a successful history as a carrier for every automobile manufacturer in the world and as a transporter of over 11 million vehicles per year, the company had reached a stalemate in its attempts to reduce product damage. The company also wanted to address the behavioral aspects of on-the-job accidents as well as customer service and quality concerns in its 121 terminal locations throughout the United States and Canada.

Solution Implemented

Behavior-based technology helped management understand why previous improvement programs had been
short-lived due to vague performance expectations,
one-time recognition for reaching short-term goals,
absence of program maintenance, and lack of employee
ownership and management support. After training in
behavioral science, supervisors and managers developed a rating system aimed at achieving 100 percent
delivery of "perfect products." Drivers and terminal
managers received regular data regarding damage
claims and costs, and every performer worked with a
matrix of pinpointed behaviors and results for ensuring
internal and external customer satisfaction. The system included safe driving behaviors, pre- and post-
inspection checklists as well as specific customer
requirements for securing, loading and tying down each
type of vehicle transported. Drivers could work through
a series of performance levels to earn tangible rewards
and social recognition for reaching measurable out-
standing performance levels. Middle management and
administrative personnel were included in the performance reward loop so that the quality initiative became
an ingrained, rather than a temporary effort. Today,
after 20 years, the behavior-based system remains
vibrant. The executive team credits the process for continued success in customer retention, quality service,
safety, the receipt of a multitude of industry awards and the
high-performing workforce that makes it all possible.

Results of Intervention

- $4 million savings in paid-out damage claims within one year

- Annual driver turnover rate of 3–4 percent compared to an industry standard of 100 percent

- Named preferred carrier for the National Automobile Transporters Association (NATA)

- Recognized by Ford Motor Company as the best damage-prevention carrier

- Received (from Toyota) the President's Award for Logistics Excellence

- Retains first and major customer after six decades

"In every location, every terminal, we have success story after success story. We're successful because we've taken the behavior-based strategy and applied it every day."

– ASSISTANT VICE PRESIDENT

"What we've got now is a system that works and pays off!"

– VICE CHAIRMAN OF THE BOARD

"This system has definitely made a positive impact on the bottom line."

– DAMAGE PREVENTION ANALYST

CASE 3

Fast Food Industry:
Reduced Employee Turnover Impacts Profits

The owner and president of 21 fast food burger franchises recognized turnover and management patterns as key issues in profitability or lack thereof. Using behavior-based technology, he identified and activated elements that transformed his holdings into profit centers with satisfied employees and fulfilled customers.

Situation

The president of 21 burger franchises knew that the average employee turnover for the industry was 300 percent. Unlike some of his store managers, he didn't view this as an unsolvable and acceptable characteristic of low-wage jobs. He calculated the time and lost productivity caused by a cycle of training, hiring, and employee attrition, then verified the negative effect it had on customer service and, ultimately, the profitability of each location. The owner described the fast food business as requiring skills in a variety of areas—manufacturing, handling and storage, merchandising, sales, accounting, product quality and consistency and customer service. Those requirements combined with the fact that the majority of employees at such establishments are young and inexperienced result in a pressure cooker work environment. This president also observed that store managers were often too dependent on the supervisors who, in turn, tended to micromanage. This resulted in managers afraid to make even minor decisions without prior approval. After learning about behavioral science at a Cambridge forum sponsored by Harvard University, he decided to put the technology to work at his franchises to optimize the critical priorities for successful management strategies, employee retention and customer satisfaction.

Solution Implemented

Using behavior-based methods for identifying critical success behaviors for supervisors, managers and employees, every store set its own performance goals fueled by positive recognition, reward and salary incentives. The changes became immediately apparent as turnover dropped by 50 percent within six months. The supervisor/manager relationships improved greatly as supervisors were coached on shaping decision-making skills for managers and for rewarding those who made competent decisions without constant oversight. The owner pinpointed and emphasized good sales behaviors over sales results, an emphasis that individual store managers at first resisted until they saw the positive and profitable results. Hiring practices changed as managers began to view potential employees as pivotal links to their own success. Today every store sets its own performance standards according to the particular location and client base. The president resists comparing performance scores between locations, preferring to let each store compete to improve its own scores measured by weighted matrices of behaviors and results focused on quality, cleanliness, productivity and customer satisfaction. Turnover continues to decline and, as the president predicted, the profits at all stores are on the upswing. Also, a culture of resignation has been replaced by one of enthusiasm.

Results of Intervention

- 50% decrease in employee turnover
- Company president states "a stratospheric performance improvement" at all 21 franchise locations
- Higher profitability for all locations
- Measurable behaviors and results presented in matrices that target behaviors affecting profitability factors developed for each restaurant site and for each employee

"When our managers interview and hire people, they're no longer just looking for a body to fill a slot. They're looking for someone who is going to be a member of their team."

– PRESIDENT AND OWNER OF
21 FAST-FOOD RESTAURANTS

CASE 4

Call Center: Customer Service

Behavior-based methods helped an international organization examine and restructure existing performance measurement systems in its customer service call center, resulting in tremendous improvements in all customer service quality indicators.

Situation

A national insurance firm faced the challenge of transitioning from a traditional indemnity organization to the fast-paced world of managed care, requiring more flexibility and proactive performance on the part of its customer care call center representatives. Existing performance measures involved a quota system for number of calls taken per hour that inspired less than lackluster interactions and resulted in employee frustration and customer irritation. Even though many detailed measures and data collection tools were in place, managers and supervisor had no idea how to alter employee interactions that were negatively affecting customer service quality measures. The center was in danger of falling short of meeting the national insurance customer service standards, or National Measurement Information Systems (NMIS) goals set annually by the corporation.

Solution Implemented

A behavior-based analysis of current performance measurement systems at the call center revealed that the quota system using *number of calls per hour* forced call center employees to choose between providing thorough and courteous customer service or meeting quota. Because of management's emphasis, meeting quota won out every time with the cost of dissatisfied, even angry customers and poor quality ratings. The center subsequently changed the measure to the percent of the workday (stated as *available time*) that representatives spent on the phone resolving customer problems and answering queries. They also used a behavioral approach to set in place a system of rotating teams for maintaining regular data-based feedback and recognition to call center employees for achieving initiatives. Available time improved by 50 percent within the first week which immediately made a profound positive impact on all interrelated NMIS customer service quality goals.

Results of Intervention

- Available time, or time spent assisting customer rose from 50% to 90% within one week of changing performance management system

- Cue time (or customer on hold) dropped from average 3 minutes to 30 seconds

- Block call (customers receive busy signal) reduced from 30% to 3%

- Abandon rate (customer disconnects) dwindled from 13% to 3%

- The customer care division met or exceeded all corporate customer service quality goals

"The change was astronomical. The improvement we've had has been miraculous."

– MANAGER OF CUSTOMER SERVICE OPERATIONS

CASE 5

Distribution Center: Accuracy and Productivity

A retail chain with 5700 stores in 24 U.S. states implemented a behavior-based management process at its newest distribution center (DC). The center quickly surpassed every performance record established in the 60-year history of the chain's six other distribution centers. Today, the retail chain is hurriedly converting the management mode of its DCs to the behavior-based technology.

Situation

This retail chain organization, established in 1939, has expanded from a single store to a multibillion-dollar business that currently operates 5700 stores and opens new locations at a rate of 600 to 700 per year. With the slogan of "A better life for everyone" the organization offers budget prices on a variety of products. Seven distribution centers across the Southeast and Midwest, each with an employee base of 500 or more, service this growing chain. Typically when a new DC opens, those employees have only 8-10 weeks to become competent at regularly and accurately servicing 600-700 retail stores. These circumstances typically result in issues of high stress, turnover, safety and accuracy (meaning the right product in the right place at the right time). In a recent opening of a new million-square-foot facility, the chain's management decided to try a behavior-based approach in an attempt to avoid the many problems inherent in such an endeavor.

Solution Implemented

The new center's managers chose to focus on recognizing the specific behaviors that would bring the performance results they desired. They involved all employees in thematic contests in which their observed behaviors were posted and graphed (anonymously) as supervisors concentrated on recognizing anyone who performed the behaviors. Within months this DC had shortened the learning curve by a full year, taking six months to meet and then surpass all performance standards set by the other centers. In fact, this DC beat the standard productivity numbers from day one. Also, because the behavioral methods are positive, attendance rates remain higher than that of any other facility and turnover is lower. Rewarding the targeted behaviors related to accuracy, productivity and cleanliness has led this DC to achieve breakthrough progress in those areas. Currently, this DC, raised on a behavioral approach, services 90 more stores than do the other DCs. Today, the parent retail chain is focused on taking this approach to the other DC operations as quickly as possible.

Results of Intervention

- With a network accuracy rate of 70%, this distribution center (the only one to use behavior-based technology) consistently achieved 90% accuracy within months of opening its doors. This will save the company at least $250,000 per year. This number does not include the sales made possible by being able to provide stores with continuous shipping. Other benefits: Exception-based "moves" of product have been cut by more than 25% and payments to vendors for incorrect receipts of purchase orders have been reduced by more than one-third.

- An average attendance rate of 95% at the behavior-based DC surpasses the 88% to 90% attendance norm at this retail chain's non-behavior-based distribution centers.

- Without emphasizing productivity numbers, the DC's shipping department quickly set a record of 42 cartons processed per man hour.

- Safety shows measurable improvement since emphasis on specific safe behaviors began. Lost time accidents are down 45% and medical-only accidents are down 22%. The client estimates that this represents a savings of $50,000 at the DC.

"We are achieving better numbers than we've ever achieved before. Our accuracy is higher; our accidents are going down; our costs are lower than they've ever been and we're setting new records on throughput activity and accuracy."

– VP LOGISTICS

CASE 6

Utility and Nuclear Power: Transforming the Culture and Thriving During Change

This power supplier to two of America's major cities had already experienced the National Regulatory Commission (NRC) shutdown of one of its three nuclear facilities, undergone a massive reorganization and survived a huge workforce reduction. Now it faced deregulation. The company needed a method for meeting the imminent requirements of higher quality, competitiveness and customer satisfaction. The management team chose a behavior-based approach as the proactive method for correcting past mistakes and facing future challenges.

Situation

When the NRC shut down one of its nuclear power facilities, citing mismanagement and industry infractions, this power supplier with 10,000 employees performed a root-cause analysis. The analysis showed the company's weaknesses were primarily those of poor employee retention, lack of accountability and failure to shape and motivate employee performance. At the same time, a rate case decision forced the utility to reduce cost and increase efficiency. The management team searched for a method to measure performance and to inspire employees to improve their own work behaviors as well as work processes. After visiting several facilities that used behavior-based technology, the team chose this approach as the engine for their multifaceted new improvement efforts.

Solution Implemented

After training a core team in behavior-based methods, the facility began training 935 management and front-line teams in the technology. The organization quickly benefited from the results of 500+ active performance pinpoints in the areas of cross-training, turnaround time on management and budget reports, nuclear facility inspection, maintenance, internal and external customer service, quality, policy and procedure development, completion and implementation of business plans, billing, meter reading and productivity. Management soon realized that it had neglected a huge resource in its people. For example, by asking for operator input regarding a purchasing decision, the transportation department saved $220,000 in direct costs and $800,000 in equipment life-cycle costs. This giant utility that had once faced over $250 million in fiscal consequences plus $1.3 million in fines for its nuclear facility failure, now used behavior-based management to reopen and revitalize the plant, receiving comments from the NRC that the revised management methods had set a new and admirable standard for the industry.

Results of Intervention

- Reopened nuclear facility averting $250 million in lost annual revenues

- Completed high quality substation with less manpower, within less time and $1 million under budget!

- Saved $220,000 in equipment purchasing costs; $800,000 in equipment life-cycle costs

- Reduced customer service backlog of 9000 to 2000 in less than a year

- Significantly reduced overtime due to higher productivity

- Accomplished a ten-fold increase in completed repair of substation breaker systems within one year (from average of 13 per month to 154 per month)

- Elevated capacity factor (percentage of time the nuclear station produces electricity) to 10% higher than industry average

- Attributed $48,000 in annual savings to less time spent preparing monthly reports in one department

"We think that the behavioral approach has transformed this company from a traditional utility to an energy services company ready to compete in the next century."

– Chairman and CEO,
Electric Company and nuclear facility

CHAPTER 17

FREQUENTLY ASKED QUESTIONS ABOUT BEHAVIOR-BASED SAFETY (BBS)

What is BBS and why should we do it?

Behavior-based safety is a process that enables managers, supervisors and frontline employees to manage any behaviors which impact safety, such as wearing protective equipment, maintaining an ergonomically correct position, and following a precise routine for performing hazardous tasks. It also includes safety support behaviors such as eliminating barriers to safe behavior, assessing and adjusting the impact of organizational systems on safety, and delivering appropriate consequences. The process is based on the science of behavior, which teaches that the most effective way to improve behavior is through a system of measurement, feedback, and positive reinforcement. Behavior-based safety processes lead to greater amounts of recognition and other positive consequences to encourage improvements over time. Behavior-based safety does not replace any existing components of an organization's safety system. Rather, it adds to that system to help organizations move to the next safety level.

What kind of results can we expect?

Behavior-based safety has a proven track record. For over 30 years, companies from all over the world have made significant safety performance improvements using behavior-based safety. A typical ADI client, with a progressive safety management process, can expect to achieve a 25 percent decrease in recordable injuries within the first year of implementation. After three to five years of implementation, behavior-based safety processes typically achieve reductions in recordable incident rates of 60 to 90 percent.

How does BBS work?

BBS is based on the science of behavior, which uses a five-step process for improving the behaviors that lead to desired results:

1) PINPOINT. Critical safe behaviors are identified, or pinpointed. These behaviors include actions of frontline workers such as putting on gloves, lifting with knees bent, or securing guards on machines. In addition, behaviors of supervisors, managers, and executives are pinpointed to include behaviors they should do for their own safety, but also, behaviors they should do to support a safe working environment.

2) MEASURE. Safe behaviors are measured, or counted to see how frequently they occur. Rather than
rely on subjective impressions of how safe people
are, BBS uses observations to collect data on safe
behaviors. Ideally, the majority of the observations are peer to peer. In some work environments, self-observation is necessary since people
work alone much or all of the time. In many cases,
supervisors and managers also do observations.

3) FEEDBACK. Once data has been collected through
observation, feedback is provided to let employees
know how they are doing. Feedback is the first
step toward helping people improve their safe
behaviors.

4) REINFORCE. Positive reinforcement is delivered
when improvements are made and goals are met
regarding safe performance. This is the most
important step in BBS—recognizing when people
are doing work in a safe manner and particularly
when they are showing improvement. Positive
reinforcement helps accelerate behavior change
and creates a positive work environment.

5) EVALUATE. After the process is up and running it
is important to check to see if everything is working as planned and if not, to go back and make
adjustments to one or more of the steps to ensure
maximum results.

How much time will this take?

The time it takes to use BBS depends on your role in the
process and the stage of the process. In the early stages,

planning and training for select groups of frontline workers and all managers and supervisors takes 1-2 days. Once this stage is completed, minimal time is required to do the day-to-day process. Observations takes only 3-5 minutes a day and managers and supervisors can expect to spend 5-10 minutes a day on the overall process. We think you'll agree that a few minutes a day toward creating a safe work environment is an investment well made.

What will be expected of me?

Everyone has a role to play in BBS. First, employees at all levels will be asked to engage in targeted safe behaviors. In addition, the process requires a group of frontline workers from each work area to act as observers. Observers spend just a few minutes each day observing and counting a few targeted behaviors. Supervisors and managers will also be asked to do observations on their own safe behaviors and on behaviors designed to help support the frontline. Finally, a small number of roles require additional time and effort (such as Steering Team membership). However, those are voluntary positions so you only have to get involved if you want to.

Is this just a policing system?
Are the observers going to be telling on peers?

Absolutely not! BBS is not about catching people doing at-risk things and reporting to management; in fact it is the opposite. BBS is about checking to see how good people already are, working with them directly to improve, and then celebrating that improvement. Observations are anonymous: that is NO NAMES are recorded. This no name/no blame observation method is simply a measurement tool that provides information about how safe the group is. Without data, we can't recognize and celebrate improvement. With BBS, group scores are collected and the group works as a team to improve on critical safe behaviors. The frontline groups, not management, handle observation sheets, so you can see it is not about "telling" on people. It is about getting an accurate picture of how the group is doing, and then working together to improve.

What's in it for me?

First and foremost, BBS will lead to an improvement in your personal safety and the safety of your peers at work. Secondly, you will be recognized for the things you and your peers currently do safely and all the improvements you make. Let's face it; safety is often a "no news is good news" issue. We usually only hear about it when something is wrong: an accident, a near miss, a poor audit score, etc. BBS teaches that if we want to optimize safety, we need to build-in systematic *positive* reinforcement for all the good things that people do for safety. Thus, you'll see a dramatic increase in the amount of recognition you and others receive for safety activities.

What are the critical success factors?

Like any other initiative (and most companies have had their share of "initiatives"), BBS depends on the people of the organization. BBS requires management support to sustain the process long term. It also requires, probably more than most other initiatives, the involvement and support of frontline employees. The people who are most at risk are those who work on the frontline. Therefore, those are the people who are targeted to manage BBS for their own work areas. This is not a "management-driven" process, but an "employee-driven" process and thus requires support and involvement at all levels. Finally, it requires a desire and commitment on the part of all employees to improve the safety of the work environment.

How is management support ensured?

Management support is a vague term. What exactly do we want managers to do to support a process? Once we answer this question, in other words, once we *pinpoint* management support behaviors, we can use the remaining four steps described earlier to make sure those behaviors happen. Through measurement of management support behaviors, feedback, and positive reinforcement we can ensure the support occurs. The management support behaviors are selected by employees at each work location and thus will vary from location to location. Some examples of management support behaviors include providing positive reinforcement for doing observations and for safe behaviors, removing obstacles to safe behavior, discussing BBS in production meetings, asking participants for feedback on the BBS process, and providing resources to frontline groups. Those behaviors are then put on a scorecard and maintained through a positive accountability system.

What will happen to our existing safety process (safety committees, safety training, etc.)? How does BBS fit in?

Maintaining a safe work environment requires a complete safety management system. BBS should be part of that system. Safety audits, training, SOP development, etc. should all continue. If anything, BBS will help you improve those other very important parts of your safety system. BBS is not a replacement for any of those other activities; it is an addition to your total system that will be customized to work well with your current safety system.

Will this have a negative impact on productivity or quality or customer service?

Actually, the opposite is more likely to happen. Because BBS is based on the scientific study of behavior, the concepts apply to any behavior . . . at work or at home. As people at all levels of the organization begin using these concepts, they quickly see applications to other parts of business such as quality and customer service. Because of the emphasis on positive reinforcement, employees are eager to use the process in other areas because they know it will lead to more recognition and reward for themselves and others. Furthermore, the focus on positive reinforcement in BBS has some great side effects, such as higher morale, lower turnover, greater involvement and an overall improved culture.

ABOUT ADI

Aubrey Daniels International (ADI) helps the world's leading businesses use the scientifically proven laws of human behavior to promote workplace practices vital to their long-term success. By developing strategies that reinforce critical work behaviors, ADI enables clients like RG&E, Dollar General, and Duke Energy to achieve and sustain consistently high levels of performance, building habits within their organization. Whether it's an individual, department or organization, ADI provides the tools and methodologies to help move people toward positive, results-driven accomplishments—performing at levels beyond expectations. Headquartered in Atlanta, the firm was founded in 1978.

SAFETY LEADERSHIP
AND CULTURE SURVEY

ADI's behavior-based Safety Leadership and Culture Survey assesses organizational and supervisory safety practices as experienced by the organization's employees. This survey gathers information on the workplace conditions, work processes, policies, procedures, and management practices that support safe behavior or that unintentionally encourage or reinforce unsafe behavior. The results from this survey help organizations identify gaps between their work environment and their safety values, while providing the feedback and accountability needed to focus busy leaders on safety.

For more information contact
Tyler Feola: 678.904.6150
or
surveys@AubreyDaniels.com

www.AdiSurveys.com